Bryophytes

*Edited by Marko S. Sabovljević
and Aneta D. Sabovljević*

Published in London, United Kingdom

IntechOpen

Supporting open minds since 2005

Bryophytes
http://dx.doi.org/10.5772/intechopen.73787
Edited by Marko S. Sabovljević and Aneta D. Sabovljević

Contributors
Preeti Chaturvedi, Dheeraj Gahtori, Elvira Baisheva, Satish Chandra, Dinesh Chandra, Arun Kumar Khajuria, Sergei Yu. Popov, Marko Sabovljević

Notice
Statements and opinions expressed in the chapters are these of the individual contributors and not necessarily those of the editors or publisher. No responsibility is accepted for the accuracy of information contained in the published chapters. The publisher assumes no responsibility for any damage or injury to persons or property arising out of the use of any materials, instructions, methods or ideas contained in the book.

First published in London, United Kingdom, 2020 by IntechOpen
IntechOpen is the global imprint of INTECHOPEN LIMITED, registered in England and Wales, registration number: 11086078, 7th floor, 10 Lower Thames Street, London, EC3R 6AF, United Kingdom
Printed in Croatia

British Library Cataloguing-in-Publication Data
A catalogue record for this book is available from the British Library

Additional hard and PDF copies can be obtained from orders@intechopen.com

Bryophytes
Edited by Marko S. Sabovljević and Aneta D. Sabovljević
p. cm.
Print ISBN 978-1-83880-144-1
Online ISBN 978-1-83880-219-6
eBook (PDF) ISBN 978-1-83880-220-2

We are IntechOpen,
the world's leading publisher of
Open Access books
Built by scientists, for scientists

4,700+
Open access books available

121,000+
International authors and editors

135M+
Downloads

151
Countries delivered to

Our authors are among the

Top 1%
most cited scientists

12.2%
Contributors from top 500 universities

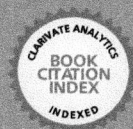

CLARIVATE ANALYTICS
BOOK
CITATION
INDEX
INDEXED

WEB OF SCIENCE™

Selection of our books indexed in the Book Citation Index
in Web of Science™ Core Collection (BKCI)

Interested in publishing with us?
Contact book.department@intechopen.com

Meet the editors

Dr. Marko S. Sabovljević is a biologist with a master's degree in Biology from the University of Belgrade, Serbia (2003) and a doctoral degree from the University of Bonn, Germany (2006). His research is dedicated to various aspects of bryophyte biology, and he has published his research in nearly 500 peer-reviewed articles. He has also authored or coauthored 11 book chapters, 189 communications, and several university and academic textbooks and monographs. He is a peer reviewer, coeditor and editorial board member for many international journals and is Editor-in-Chief of Botanica Serbica. Currently, Dr. Sabovljević is a fulltime professor of Ecology, Biogeography and Environmental Sciences in the Department of Plant Ecology and Phytogeography, Institute of Botany and Botanical Garden, Faculty of Biology, University of Belgrade, Serbia.

Dr. Aneta D. Sabovljević is a biologist with a master's degree in Biology from the University of Belgrade, Serbia (2003) and a doctoral degree from the University of Belgrade, Serbia (2007). She has a rich national and international research career, and is interested in studying bryophytes both in the lab and in the field. She has mostly experimental and analytical approaches to bryophyte biology. Dr. Sabovljević has published her research results in more than 100 peer-reviewed articles. She is a peer reviewer for many national and international scientific journals. Currently, she is Associate Professor of Physiology and Molecular Biology of Plants in the Department of Plant Physiology, Institute of Botany and Botanical Garden, Faculty of Biology, University of Belgrade, Serbia.

Contents

Preface

Bryophytes are a group of plants with around 22,500 species worldwide. It is a diverse group that includes many subgroups that are phylogenetically rather different from one another, although they share a common and unique life cycle. There are marked differences between organisms in this group, including various life strategies, life forms, survival and dispersal possibilities, chemical constituents, biochemical patterns, physiological mechanisms and ecological responses. Indeed, there is such an immense diversity that we are still far from having satisfactory knowledge of bryophytes. Even if we study more classical disciplines such as biogeography, there is still much we don't know about bryophytes and their potential applications.

In this book, we offer new views, novel data and various approaches to bryophyte sciences. Chapters cover distribution changes of *Sphagna* moss, particularly *Cuspidata*, in the East European Plain and Eastern Fennoscandia, bryophyte diversity within the forests of the Southern Urals (Russia), the antioxidant potential of bryophytes, and the role of ohioensins, which are chemical constituents of some bryophyte species, in curing disease.

The editors acknowledge the contributing authors as well as the entire IntechOpen publishing team for all their attention and support, without which this book would not be possible.

Dr. Marko S. Sabovljević
Full Professor in Ecology, Biogeography and Environmental Sciences,
Faculty of Biology,
Department of Plant Ecology and Phytogeography,
Institute of Botany and Botanical Garden,
University of Belgrade,
Serbia

Dr. Aneta D. Sabovljević
Associate Professor in Physiology and Molecular Biology of Plants,
Faculty of Biology,
Department of Plant Physiology,
Institute of Botany and Botanical Garden,
University of Belgrade,
Serbia

Chapter 1

Introductory Chapter: Bryophytes 2020

Marko S. Sabovljević and Aneta D. Sabovljević

1. Bryophytes

Bryophytes, amphibians within the plant kingdom, were among the first plants that lived in the water and settled in terrestrial environment. They faced such a harsh and bare environment they needed to cope with; but on the other hand, these facts gave opportunities to diversify so much. Nowadays, they count between 18,000 and 23,000 extant species [1]. However, one can estimate many undescribed species that appear with classical and molecular approaches, and every year many new species for science are reported, even from Europe [2] that is bryologically among the best investigated world areas. Classification of such a huge diversity as bryophytes is the matter of discussions leading to treating them as one division or alternatively three or more within the common subkingdom of plants named Bryobiotina. Irrelevant of classification level, the three widely accepted group can be considered within bryophytes: mosses (11,000–13,000 species), liverworts (7000–9000 species), and hornworts (200–250 species).

During the decades of accumulation on the knowledge on bryophytes, bryophyte science developed and changes our views on the three group relationships among themselves and with other plants and algae. One of the latest views is that mosses and liverworts can be considered as Setaphyta while hornworts (Anthocerotophyta) seems to be near vascular plants [3].

Nevertheless, these groups share common ancestor, as well as a number of biological and ecological traits. Even though some lignin-like compounds are discovered to be present in some species, they do not produce lignin which preclude them to develop into huge forms due to the absence of mechanical body support. They are all rather small plants, even some species can reach few decimeters in height. The life cycle of bryophytes is dominated by haploid gametophytes, while diploid sporophyte has even shorter appearance during the sexual reproduction.

The strong cuticles are absent unabling them to keep body water balance. This means they are water dependent from the water balance in the immediate environment, i.e., poikilohydric. Thus, many of bryophytes can go to anabiosis which means drying out during dry period and once the wet period is back they can quickly resume their metabolism with no damage. This is why many species belong also to the group of unrelated vegetation representatives called resurrection plants.

The lack of cuticles, whole body collecting water from the atmosphere, and the absence of rootlike structure for soaking (rhizoids have mainly anchoring function) are some of the characteristics which makes them good bioindicator species quickly reacting to small changes in proximate environment. So, bryophytes possess a huge potential in specific sensitivity also due to their relations to specific microhabitats and proximate ecological conditions, and they greatly exceed the sensitivity of vascular plants to pollutants or in general environmental changes.

Bryophytes inhabit almost all ecosystems on the Earth. There are no representatives in the seas, but there are representatives in the brackish waters or moderate salt environment [4]. Though, over the times we learn that they have significant roles within the communities they live in. A range of significant ecological functions of bryophytes are huge and vary from biomes to biomes, but the general ones include water retention (acting like a huge sponge; they play a significant role in water balancing in the ecosystems), carbon sequestration (i.e., carbon locking by peatland dominated by peat mosses), or biotic interaction with other organisms (e.g., providing shelters, acting as a seed bed, or representing habitat *per se* for many other organisms).

The geographical ranges of bryophytes are wider than those of vascular plants [5]. This is due to the long-distance dispersal of small spores, huge survival rate or diaspores during transportation, and settling specific microhabitats over the huge areas. In general, we can say they are ubiquitous, since species can be found in dry desert to the underwater deep in freshwater lakes and from the sea level till the top of the highest mountains surviving even under the long laying ice. Their nutrient supply is over the whole body surface coming from precipitates. Drying out, i.e., suspending physiological activities, versus rewetting, i.e., establishing back normal life function, can occur on a daily basis (e.g., *Grimmia* and *Schistidium* that live on exposed rocks), or they can survive longer periods of inactivity upon dehydration (many members of Pottiaceae).

Most of the bryophyte species are rather less competitive to resources in the environment than vascular plants. Thus, they have a wide range of distinctive feature to survive including, beside the abovementioned, life forms and life strategies. Also dispersal and propagation can be through various vectors both biotic (e.g., birds, snails, mammals) and abiotic (e.g., wind, watercourses), and apart from spores that are produced sexually, diaspores can be produced on rhizoids (e.g., *Bryum*), stem tips (*Aulacomnium*), or on the leaves (*Pohlia*, *Orthotrichum*, etc.) for vegetative spread. Even parts of the whole bodies can serve for this purpose. Such an efficient possibility for wide dispersion and long viability of spores and diaspores enables them to rich long distances, and this is why the endemism is rather low compared to vascular plants. Some authors stated that 10% among European bryophytes express the endemic characters compared to 28% of tracheophytes. On the other hand, the discontinuous ranges and disjunctions are very high among bryophytes [5] as a consequence of very efficient spread and the microhabitat importance for new population establishment.

Among the interesting features of bryophytes, being an ecological indicator should not be passed by (e.g., [6–8]). Many species occur on specific pH of the substrate, or indicate by appearance air quality. Additionally, some species are so well adapted to substrate and nutrients coming from dissolved substrate by precipitation that they can indicate the presence of salt (e.g., *Entosthodon hungaricus*) [9] or minerals (Pb, *Ditrichum plumbicola*; Cu, *Scopelophila cataractae*) [10] and are strict to such a region.

Fast-spreading protonemal growth in a short suitable period can stabilize the soil surface, preventing erosion. Also, they are pioneer colonizers and stabilizers of bare surface, enriching the ecosystems and producing a suitable habitats for further settlers in successional phases of ecosystem changes/development. Spreading colonies on rocks, they initiate soil establishment and participate in protosoil production cohabiting with cyanobacteria, playing an important role in nitrogen fixation, i.e., enabling colonization for other plants and organisms.

Bryophytes have no huge commercial values, at present. These values come from a huge number of both biotic and abiotic interactions as well as their peculiarities (e.g., [6, 11, 12]). However, they have huge applicative potential, which is lately noticed but still neglected compared to vascular plants.

Chemical constituents of bryophytes attract lately very much attention since many new to science and rare or modified known compounds are discovered in different species [13–16]. Also, bioactivities of extracts or target compounds are promising for bio-industrial products such as biopharmaceutical, biopesticides, biorapelents, or cures. Since all these products are environmentally friendly, new biotechnological processes with bryophytes are needed to be established to get to the point when wide use can be done. The treat to some modern diseases like AIDS and different cancer types and even new antibiotics are possible to develop from bryophytes [17–19]. Huge potential of bryophytes are seen by cosmetic industry as well. The problems remain the small biomass in nature for such a project, hard identification, monoculturing, cohabitation, and interfusion with many other organisms. There are steps forward to establish bryo-reactors with selected species to overcome these problems, but still clean start material is needed to do so. Therefore, the axenic and in vitro establishment of target taxa is necessary. This is not an easy task, having in mind that many species are hardly available and not in a proper developmental stage, and also due to lack of cuticles, one-cell thalli layers that unable or hardening surface sterilization without killing target material as well. Additional problem can be endophytic cohabitants.

Though, there are many advantages in bryophytes. For example, easy gene targeting and high rate of homologous recombination are the main pathways for transforming DNA to incorporate in moss genome [5, 20]. This is surely true for the model moss whose genome is completely sequenced, namely, *Physcomitrella patens*. It is widely studied and exhibits high frequencies of gene targeting. DNA constructs with sequences homologous to genomic loci can transform moss rather easy. The outcome then is the organism with targeted gene replacement resulting from homologous recombination although untargeted integration at nonhomologous sites can also occur, but at a significantly lower frequency which can be easily eliminated.

Since, these organisms are rather microhabitat dependent and sensitive to environmental changes, large-scale harvesting and impulsive climate change can cause both diversity and biomass loss not only damaging bryophytes *per se* but the global ecosystem as well. Thus, protection and conservation for the bryophytes are urgently needed in a quick-changing world [21]. Many governments and conservationist have already done a lot in legislative, giving priority to highly threatened species, i.e., applying passive measures for the well-being of mosses, liverworts, and hornworts or habitats they live in. However, it seems these are not enough, and the decrease in populations, even species loss, is taking part. Therefore active conservation measures are needed: species propagation, species reintroduction, habitat management, and constant monitoring [22–27]. The emerging field of conservation biology, namely, conservation physiology, is therefore needed to learn in experimental both laboratory and field conditions, those what is essential on species biology prior to decision which measures will be applied for good species conservation and loss prevention apart from legal measure.

And again, in vitro establishment, studies, and propagation arise as problem solutions in maintaining ex situ collections and preparing material for release to the wild [25, 27]. The reviving of material stored in herbarium is sometimes possible [26], but in most cases good green material is needed which mostly is not the case.

Many problems in different fields of bryophyte sciences remain to be solved, and many phenomena remain to be uncovered, although over the past century many knowledge on bryophyte biology were accumulated. However, in 2020 we still need both to spread among known fact searching for overlooked and to go deeper beyond the point that has been reached up to date.

Author details

Marko S. Sabovljević* and Aneta D. Sabovljević
Faculty of Biology, Institute of Botany and Botanical Garden, University of
Belgrade, Belgrade, Serbia

*Address all correspondence to: marko@bio.bg.ac.rs

IntechOpen

References

[1] Villareal JC, Cargill DC, Hagborg A, Soderstrom S, Renzgalia KS. A synthesis of hornwort diversity: Patterns, causes and future work. Phytotaxa. 2010;**9**:150-166

[2] Pocs T, Sabovljevic M, Puche F, Segarra Moragues JG, Gimeno C, Kürschner H. Crossidium laxefilamentosum Frey and Kürschner (Pottiaceae), new to Europe and to North Africa. Studies on the cryptogamic vegetation on loess clifs, VII. Journal of Bryology. 2004;**26**:113-124

[3] Puttick MN, Morris JL, Williams TA, Cox CJ, Edwards D, Kenrick P, et al. The interrelationships of land plants and the nature of ancestral embryophyte. Current Biology. 2018;**28**:733-745

[4] Ćosić M, Vujičić MM, Sabovljevic MS, Sabovljevic AD. What do we know on salt stress in bryophytes? Plant Biosystems. 2019;**153**:478-489

[5] Frahm JP. The phytogeography of European bryophytes. Botanica Serbica. 2012;**36**:23-36

[6] Sabovljevic MS, Weidinger ML, Sabovljevic AD, Adlassing W, Lang I. Is binding patterns of Zn(II) equal in different bryophytes? Microscopy and Microanalysis. 2018;**24**:69-74

[7] Anicic Urosevic M, Vukovic G, Jovanovic P, Vujicic M, Sabovljevic A, Sabovljevic M, et al. Urban background of air pollution: Evaluation through moss bag biomonitoring of trace elements in botanical garden. Urban Forestry & Urban Greening. 2017;**25**:1-10

[8] Berisha S, Skudnik M, Vilhar U, Sabovljevic M, Zavadlav S, Jeran Z. Trace elements and nitrogen content in naturally growing moss Hypnum cupressiforme in urban and peri-urban forests of the municipality of Ljubljana

(Slovenia). Environmental Science and Pollution Research. 2017;**24**:4517-4527

[9] Sabovljevic MS, Nikolic N, Vujicic M, Sinzar-Sekulic J, Pantovic J, Papp B, et al. Ecology, distribution, propagation in vitro, ex situ conservation and native population strengthening of rare and threatened halophyte moss *Entosthodon hungaricus* in Serbia. Wulfenia. 2018;**25**:117-130

[10] Stankovic J, Sabovljevic AD, Sabovljevic MS. Bryophytes and heavy metals: A review. Acta Botanica Croatica. 2018;**77**:109-118

[11] Ručová D, Goga M, Sabovljevic M, Vilková M, Petruľová V, Bačkor M. Insights into physiological responses of mosses *Physcomitrella patens* and *Pohlia drummondii* to lichen secondary metabolites. Protoplsma. 2019;**256**:1585-1595

[12] Goga M, Ručova D, Kolarcik V, Sabovljevic M, Bačkor M, Lang I. Usnic acid, as a biotic factor, changes the ploidy level in mosses. Ecology and Evolution. 2018;**8**:2781-2787

[13] Pejin B, Vujisic L, Sabovljevic M, Tesevic V, Vajs V. The moss *Mnium hornum*, a promising source of arachidonic acid. Chemistry of Natural Compounds. 2012;**48**:120-121

[14] Pejin B, Bianco A, Newmaster S, Sabovljevic M, Vujisic L, Tesevic V, et al. Fatty acids of Rhodobryum ontariense (Bryaceae). Natural Product Research. 2012;**26**:696-702

[15] Pejin B, Sabovljevic M, Tesevic V, Vajs V. Further study of fructooligo-saccharides of Rhodobryum ontariense. Cryptogamie Bryologie. 2012;**33**:191-196

[16] Pejin B, Iodice C, Tommonaro G, Sabovljevic M, Bianco A, Tesevic V, et al. Sugar composition of the moss

Rhodobryum ontariense (Kindb.) Kindb. Natural Product Research. 2012;**26**:209-215

[17] Sabovljevic MS, Vujicic M, Wang X, Garraffo M, Bewley CA, Sabovljevic A. Production of the macrocyclic bis-bibenzyls in axenically farmed and wild liverwort Marchantia polymorpha L. subsp. ruderalis Bischl. Et Boisselier. Plant Biosystems. 2017;**151**:414-418

[18] Sabovljevic MS, Sabovljevic AD, Ikram NKK, Peramuna A, Bae H, Simonsen HT. Bryophytes – An emerging source for herbal remedies and chemical production. Plant Genetic Resources. 2016;**14**:314-327

[19] Sabovljevic M, Bijelovic A, Grubisic D. Bryophytes as a potential source of medicinal compounds. Lekovite Sirovine. 2001;**21**:17-29

[20] Kamisugi Y, Schlink K, Rensing SA, Schween G, von Stackelberg M, Cuming AC, et al. The mechanism of gene targeting in Physcomitrella patens: Homologous recombination, concatenation and multiple integration. Nucleic Acids Research. 2006;**34**:6205-6214

[21] Hodgetts N, Calix M, Englefield E, Fettes N, Garcia Criado M, Patin L, et al. A Miniature World in Decline: European Red List of Mosses, Liverworts and Hornworts. Brussels, Belgium: IUCN; 2019

[22] Sabovljevic MS, Segarra-Moragues JG, Puche F, Vujicic M, Cogoni A, Sabovljevic A. Eco-physiological and biotechnological approach to conservation of the world-wide rare and endangered aquatic liverwort Riella helicophylla (Bory et Mont.) Mont. Acta Botanica Croatica. 2016;**75**:194-198

[23] Sabovljevic M, Papp B, Sabovljevic A, Vujicic M, Szurdoki E, Segarra-Moragues JG. In vitro micropropagation of rare and endangered moss *Enthostodon hungaricus* (Funariaceae). Bioscience Journal. 2012;**28**:632-640

[24] Sabovljevic M, Ganeva A, Tsakiri E, Stefanut S. Bryology and bryophyte protection in the South-Eastern Europe. Biological Conservation. 2001;**101**:73-84

[25] Sabovljevic M, Vujicic M, Pantovic J, Sabovljevic A. Bryophyte conservation biology: In vitro approach to the ex situ conservation of bryophytes from Europe. Plant Biosystems. 2014;**148**:857-868

[26] Vujicic M, Sabovljevic A, Sinzar-Sekulic J, Skoric M, Sabovljevic M. In vitro development of the rare and endangered moss Molendoa hornschuchiana (hook.) Lindb. Ex Limpr. (Pottiaceae, Bryophyta). HortScience. 2012;**47**:84-87

[27] Rowntree JK, Pressel S, Ramsay MM, Sabovljevic A, Sabovljevic M. In vitro conservation of European bryophytes. In Vitro Cellular & Developmental Biology–Plant. 2011;**47**:55-64

Chapter 2

Bryophyte Diversity in the Forests of the Southern Urals

Elvira Baisheva, Pavel Shirokikh and Vasiliy Martynenko

Abstract

Mountain forest monitoring is closely related to the survey of bryophyte species since it is there that these organisms are common and show very specialized ecological niches. This work is aimed to show how bryophyte richness and taxonomic and ecological categories differ in the various types of indigenous forests in the Southern Ural Mountains. The distribution of bryophytes in mountain forests of the Southern Urals was examined at about 1700 sample plots. Frequency and abundance patterns suggested that species richness, taxonomic distribution, and substrate group distribution are mostly determined by the forest type. According of bryological data, the forest associations characterized by high diversity and concentration of rare species were identified. This is mainly tall herb spruce-fir and mixed forests. The proportion of rare species in these forests is about 9%, including a significant number relicts both of European and Asian origins. The sites of these forests are most valuable for nature conservation and should be protected.

Keywords: bryophytes, biodiversity, forest vegetation, the Southern Urals, syntaxonomy, nature protection

1. Introduction

The Southern Urals is the southern part of Ural Mountains that is also called the Urals, or Ural, a mountain range in west-central Russia which borders with East European Plain to the west and West Siberia to the east. The vegetation in the Southern Urals is characterized by high diversity. It depends on a number of regional factors: unique geographic position between Europe and Asia; mountains being an important climatic boundary, causing significant differences in vegetation of the western and eastern slopes; and the absence of Pleistocene glaciation, which allowed to preserve the ancient elements of the vegetation [1].

Currently, the forests cover about 80% of the Southern Urals [2]. The modern altitudinal forest belt and floristic composition of current Urals forests were formed over the last 4500 years during the subboreal and sub-Atlantic periods of the Holocene. At the same time, the Ural ridge became a natural physical and geographical boundary for the ranges of many nemoral species due to the increasing of climate continentality from west to east [3]. The Ural ridge is a natural barrier to the path of moist and warm Atlantic air masses. For this reason, the climate on the western Ural's slope is humid and warm; it is more favorable for the deciduous forests. On the eastern slope, the climate is more continental, which led to the dominance of hemiboreal pine-birch and larch forests of West Siberian type and steppe communities. In the middle of the central elevated part of the Southern Urals, the spruce-fir and mixed broad-leaved and dark coniferous forests are widespread. Thus, the Southern Urals is a contact area of three types of forest vegetation, that is, (1) East European broad-leaved deciduous forests, (2) mountain taiga, and (3) hemiboreal Siberian pine, larch, and birch forests. The presence of the species from three complexes, nemoral, boreal, and hemiboreal, significantly increases species richness of Ural forests [3–5].

In comparison with North European forests, where agriculture started to reduce in the forest area 3000–5000 years ago [6], the history of forest exploitation in the Southern Urals is relatively short: the large-scale felling has been carried out here for 300 years. Nevertheless, intensive cutting of the mountain forests in this area has brought about a highly mosaic structure of vegetation represented by various types of forests at different stages of regeneration and age succession [7]. Also, heavily exploitation of mountain forests causes a major part of their biodiversity value to be lost.

One of the important challenges of forest ecology is reconstruction of the picture of the original species composition and structure of the forests (in relation to overall biodiversity, space, age, thickness of trees, composition of understory vegetation, etc.). The preserved parts of indigenous forests represent an important base for knowledge about the original diversity and structure of forests of a particular region. The identification of the most vulnerable and valuable forest sites in terms of the conservation of forest biodiversity is an important conservation challenge which attracts the attention of specialists from different fields. There are several approaches to the selection of these areas, which differ by their significance, sizes, and level of conservation. In the vegetation science, there are many categories of "etalon" forest communities: forests of high conservation value, intact forest landscapes (IFA), biologically valuable forests (IVF), etc. [8]. Despite the differences in the interpretation of the terms "indigenous," "old-growth," or "intact" forests, most of the researches keep in mind forest communities that have developed over a long period of time (comparable to the maximum biological age of tree species or exceeding this age) without or with minimal human impact [9]. In our research, we identify these forests as "relatively indigenous," or "indigenous," keeping in mind that at least some of them were established in the sites that were more or less human-modified in the past. It should be noted that many types of indigenous forest communities of the Southern Urals have been preserved only by small fragments in areas difficult to access for logging or in specially protected natural areas [10, 11].

Bryophytes represent a significant component of overall forest diversity and play important roles in ecosystem processes and functions [12], but in European Russia the bryophyte diversity in particular types of forest is not well studied [13–15].

In the last three decades, we are specifically interested in the revealing of bryophyte diversity in the forests of the Southern Urals and the presence of rare species in the different forest types of the region. In particular, we intended to answer the following questions: (1) Do bryophyte diversity vary in relation to environmental variables linked to different forests types? (2) Are rare and endangered species mainly associated with particular forest types? (3) What differences exist between bryophyte diversity in indigenous and secondary forest communities established after felling?

2. Study area and characteristic of investigated forest types

The research was performed between 1991 and 2017 in the territory of the Republic of Bashkortostan and some adjacent areas of Chelyabinsk Region (Russian Federation). The climate of study area is moderately continental, with relatively

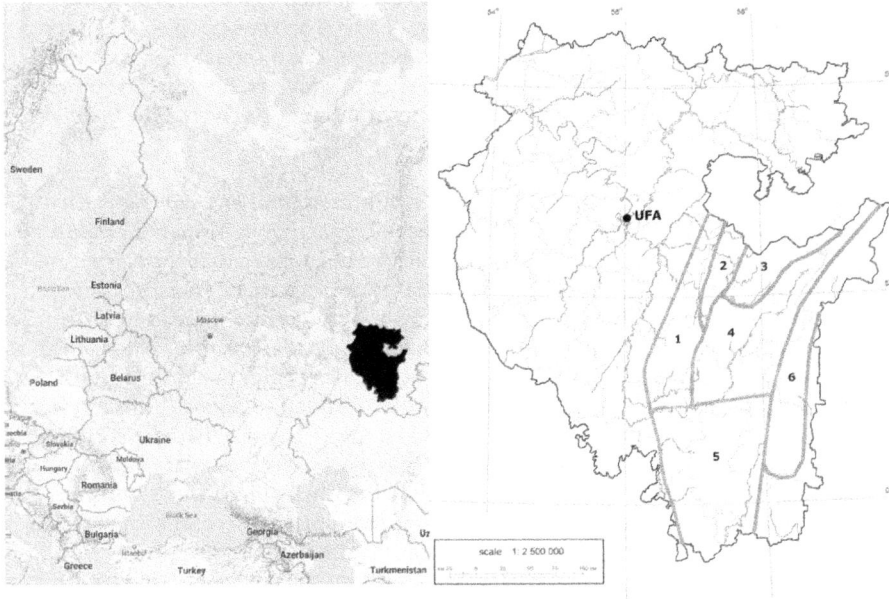

Figure 1.
Map of study area and the districts of natural zoning of the Southern Urals according to the scheme of A. Muldashev [16].

warm summer and long cold winter. The average annual temperature is +0.5 to +2.0; the mean temperature in January is −15.5 to −17°C and in July +16.5 to +17.5°C. The mean annual precipitation is 500–700 mm. The frost-free period is 50–90 days; the mean snow depth is 60–75 cm [2].

According to natural zonation of the Republic of Bashkortostan (**Figure 1**), the study area belongs to six districts [16]:

1. *The district of mountain broad-leaved forests of the western slope of the Southern Urals* includes Bash-Alatau, Takaty, Kyrybujan, Ulutau, Alatau, Kolu, Kanchak, and Kibiz Ranges extended mostly in the meridional direction. The relief is presented by low and middle mountains dissected by deep river canyons. The soil cover consists mostly of mountain forest gray and dark gray soils. The watersheds and mountain slopes are covered by broad-leaved forests. On the southern slopes, the oak forests are more common, and the northern slopes are covered mostly by maple woodlands. The lime forests grow in the low parts of gentle slopes of different exposure. In the river valleys, the floodplain forests with elm, alder, and bird cherry are widespread. In the valleys of Zilim and Nugush Rivers, the fragments of old-growth pine and spruce forests are preserved. The largest areas are occupied by secondary lime, birch and aspen forests, and the meadows. The territory is sparsely populated.

2. *Zil'merdak District of mixed deciduous and dark coniferous fir and spruce forests* of the middle part of the Southern Urals includes Zil'merdak Range and adjacent mountains. The relief is hills, terrain, and middle mountains. The soil cover consists mostly of mountain forest light gray soils. In the past, the spruce-fir forests with an admixture of lime, maple, elm, and oak, as well

as pine broad-leaved forests, were widely distributed here. Currently, the indigenous vegetation is preserved in small areas, on steep slopes, and in the water-protected zone of the Inzer River. Most of the district's area is covered by secondary birch, aspen, and mixed deciduous forests. Steppe communities occur on steep rocky slopes in the canyons of mountains rivers. The area is sparsely populated.

3. *Yamantau District of dark coniferous fir-spruce forests and highland vegetation* comprises the largest peaks of the Southern Urals, that is, Yamantau Mountain (1638 m.a.s.l.) and Iremel Mountain (1586 m.a.s.l.), as well as the Nary, Mashak, and Belyagush Ranges. The relief is ridges with deep depressions. Mountain sod-podzol, mountain gray forest, and mountain meadow soils are widespread. Before the forest exploitation, at the altitude of (900) 1000–1100 m.a.s.l., the mountain slopes have been covered by spruce-fir forests, sometimes with an admixture of bush form of linden. Also, the pine and larch forests were quite common in the upper and lower forest belts, respectively. The present-day vegetation is represented by different secondary successional communities, appeared after felling, mostly by birch woodlands. In the subalpine belt, up to 1150–1250 m.a.s.l., the spruce, fir, and birch (*Betula czerepanovii* Orlova) open woodlands, as well as tall-herb alpine meadows, are widespread. The tops of some highest mountains are covered by mountain tundra. In river valleys bogged birch-spruce forests are quite common. The area is very sparsely populated.

4. *The district of light coniferous pine and larch forests in the central part of the Southern Urals* includes the Jurmatau, Belyatur, Shatak and Kraka Ranges (800–1000 m.a.s.l.) and part of the Uraltau Range. Relief is ridges and intermountain depressions. Mountain forest gray soils are predominated. The pine forests are very common, somewhere, for example, on Kraka and Jurmatau Ranges; the large sites of larch forests bordered with petrophyte steppe communities are presented. Most of indigenous forests are replaced by secondary birch and aspen woodlands. At the altitude 850–900 m.a.s.l. near the upper border of forest belt appear larch-birch scarce woodlands and the meadows. The area is well developed and relatively densely populated.

5. *The district of forest and forest-steppe in the Zilair Plateau* has the relief consisting of the low ridges and hills with deeply embedded river valleys. The different types of chernozems and gray and dark gray forest soils are common here. Western part of plateau is covered with oak forests that are replaced with pine broad-leaved forests in an eastward direction. The coniferous forests are mostly cut down and replaced with birch woodlands. The area is fairly sparsely populated.

6. *The forest-steppe district of eastern slope of the Southern Urals* includes Krykty, Kurkak, and Irendyk Ranges and adjacent foothills. The relief is low ridges and hills. In the depressions, the lakes and mires are quite common. The soil cover consists mostly of mountain gray forest soils and leached chernozems. In the past, the upper parts of the ridges were occupied by pine, larch, and birch forests; nowadays vegetation is represented mainly by the birch forest-steppe. The region is relatively densely populated.

The main forest trees of the Southern Urals are *Picea obovata* Ledeb., *Abies sibirica* Ledeb., *Pinus sylvestris* L., *Larix sibirica* Ledeb., *Tilia cordata* Mill., *Acer platanoides* L., *Quercus robur* L., *Ulmus glabra* Huds., *Betula pendula* Roth., *Betula pubescens* Ehrh., *Populus tremula* L., and *Alnus incana* (L.) Moench and *Padus avium* Mill. The floodplain forests are formed by *Populus nigra* L., *Populus alba* L., and *Salix* spp.; also paludified floodplain forests with predominance of *Alnus glutinosa* (L.) Gaertn. are occasionally found.

Vegetation classification of the indigenous forests of the Southern Urals consistent with the Braun-Blanquet approach [17] allowed to separate 42 associations which belonged to 10 sub-alliances, 9 alliances, 4 orders, and 4 classes [10]. In total, bryophyte richness was studied in 11 types of forest communities (F1–F11) separated at the level of alliances and sub-alliances. The syntaxonomical position of investigated forest types is given below:

Class CARPINO-FAGETEA SYLVATICAE Jakucs ex Passarge 1968 (syn. Querco-Fagetea Br.-Bl. et Vlieger in Vlieger 1937)—mesic deciduous and mixed coniferous-broadleaved forests of temperate Europe, the Caucasus and Southern Siberia [18].

Order FAGETALIA SYLVATICAE Pawłowski 1928.

F1—Alliance Alnion incanae Pawłowski et al. 1928.

F2—Alliance Lathyro pisiformis-Quercion roboris Solomeshch et Grigoriev in Willner et al. 2015.

Alliance Aconito lycoctoni-Tilion cordatae Solomeshch et Grigoriev in Willner et al. 2016.

F3—Sub-alliance Aconito septentrionalis-Tilienion cordatae Martynenko 2009 prov.

F4—Sub-alliance Tilio cordatae-Pinenion sylvestris Martynenko et Shirokikh in Martynenko 2009 prov.

Class ASARO EUROPAEI-ABIETETEA SIBIRICAE Ermakov, Mucina et Zhitlukhina in Willner et al. 2016—cool-temperate coniferous and mixed montane forests with nemoral and hemiboreal floristic elements of the Southern Urals and Southern Siberia [18].

Order ABIETETALIA SIBIRICAE (Ermakov in Ermakov et al. 2000) Ermakov 2006.

F5—Alliance Aconito septentrionalis-Piceion obovatae Solomeshch, Grigoriev, Khaziakhmetov et Baisheva in Martynenko et al. 2008.

Class VACCINIO-PICEETEA Br.-Bl. in Br.-Bl. et al. 1939—holarctic coniferous forests on oligotrophic and leached soils in the boreal zone and at high-altitudes of mountains in the nemoral zone of Eurasia [18].

Order PICEETALIA EXCELSAE Pawłowski et al. 1928.

Alliance Piceion excelsae Pawłowski et al. 1928.

F6—Sub-alliance Atrageno sibiricae-Piceenion obovatae Zaugolnova et al. 2009.

F7—Sub-alliance Eu-Piceenion abietis K.-Lund 1981.

Order PINETALIA SYLVESTRIS Oberd. 1957.

F8—Alliance Dicrano-Pinion sylvestris (Libbert 1933) W. Matuszkiewicz 1962.

Class BRACHYPODIO PINNATI-BETULETEA PENDULAE Ermakov et al. 1991 hemiboreal pine and birch-pine herb-rich open forests on fertile soils of the Southern Urals and Southern Siberia [18].

Order CHAMAECYTISO RUTHENICI-PINETALIA SYLVESTRIS Solomeshch et Ermakov in Ermakov et al. 2000.

F9—Alliance Caragano fruticis-Pinion sylvestris Solomeshch et al. 2002.

F10—Alliance Veronico teucrii-Pinion sylvestris Ermakov et Solomeshch in Ermakov et al. 2000.

F11—Alliance Trollio europaea-Pinion sylvestris Fedorov in Ermakov et al. 2000.

Short characteristic of the investigated forest types is given in **Table 1**.

Forest type	Number of relevés	Elevation	Districts	Description	Cover of moss layer, %
F1	96	200–700	1–6	Floodplain gray alder forests	1–2
F2	115	250–720	1, 3, 5	Thermophytic oak forests	1–10
F3	287	220–630	1, 4, 5	Mesic linden-maple-oak forests	<1
F4	139	350–550	1, 4, 5	Mesic pine-linden-oak forests	<1
F5	195	500–1300	1–5	Spruce-fir and mixed forests with nemoral herb layer	1–20
F6	132	400–1400	1–5	Spruce-fir green moss—tall-forb forest	55–80
F7	52	450–1300	3	Spruce-fir green moss forest	75
F8	120	350–1100	2–4	Pine green moss forest	20–70
F9	88	400–900	1, 3–5	Hemiboreal xero-mesophytic pine and pine-larch forests	5–10
F10	101	400–750	4–6	Hemiboreal birch and birch-pine forests	1–5
F11	179	450–950	1–6	Mesophytic hemiboreal birch and larch pine forests	3–15

Table 1.
Characteristic of the forest types studied (F1–F11) in the Southern Ural Mountains, Russia.

3. Field sampling

In about 400 localities, we have laid 1700 sample plots with area of 400 m^2 (usually, not <10 plots for each of forest association). In each plot, the phytosociological relevés, following the Braun-Blanquet method [17], were carried out. All bryophytes growing on various substrates (soil, tree trunks, rotten wood, rock outcrops, etc.) were described.

An estimate of the abundance of forest floor bryophyte species was carried out on the basis of the Braun-Blanquet scale: r, a single species is encountered (the cover is insignificant); +, the species is rare and has a small projective cover up to 1%; 1, the projective cover of species is 1–5%; 2, projective cover is 6–25%; 3, projective cover is 26–50%; 4, projective cover is 51–75%; and 5, projective cover is more than 75%. In synoptic **Table 4**, the following scale of constancy was used: r, the species was encountered in <5% of the relevés; +, in 6–10% of relevés; I, in 11–20% of relevés; II, 21–40% of relevés; III, 41–60% of relevés; IV, in 61–80% of relevés; V, in 81–100% of relevés. Classification of forest vegetation was conducted according to the Braun-Blanquet approach using the TURBOVEG database [19] and program JUICE [20]. Similarity of species diversity between different forest types was calculated using the Jaccard index with program IBIS [21].

Additionally, collections of bryophytes were made outside the sample plots of phytosociological relevés (on forest roads, banks of streams, on bare soil near the roots of fallen trees, etc.). These data were used for compiling of checklist of bryophytes revealed in the forests of the Southern Urals. The specimens are kept in the Herbarium of Ufa, Institute of Biology—Subdivision of the Ufa Federal Research Centre of the Russian Academy of Sciences (UFA), and partly in Herbarium of the Main Botanical Garden of Russian Academy of Sciences (MHA).

4. The species richness and substrate groups of bryophytes in different types of forest communities

Forest areas of the Southern Urals have a rich bryophyte flora, presenting 286 species of Bryophyta distributed across 124 genera and 44 families and 58 species of Marchantiophyta from 33 genera and 23 families. This represents 75 and 62% of the moss and liverwort richness of the Republic of Bashkortostan, respectively. High proportions of mountain forest species in the total bryophyte flora of the Bashkortostan stress the great importance of forest protection for preserving biodiversity in the republic.

The predominant families, in terms of the number of species, are Brachytheciaceae (26 species), Dicranaceae (23), Grimmiaceae (20), Amblystegiaceae (20), Sphagnaceae (18), Bryaceae (17), Mniaceae (16), Pottiaceae (14), Polytrichaceae (11), and Pylaisiaceae (12), and the leading genera are *Sphagnum* (18 species), *Dicranum* (18), *Bryum* (15), *Brachythecium* (11), *Pohlia* (10), *Schistidium* (10), *Orthotrichum* (8), *Grimmia* (7), and *Plagiomnium* (7).

Some species richness indicators within studied forest types are given in **Table 2**. The mosses have the leading position in the bryophyte flora of all investigated forests where they held 78–93% of the total species richness within different forest types (**Table 2**). The participation of liverworts is less significant; their richness increases only in the forests with dominance of dark coniferous trees (F5–F7).

The highest species richness in the dark coniferous and mixed tall-herb forests (types F5 and F6) can be explained, mainly, by the relatively high humidity in their habitats, as well as high diversity of epixylic and ground species in these communities. The lowest bryophyte diversity was revealed in the pine, birch, and larch forests (F7, F10, and F11) that developed in the more continental climate conditions of eastern slope of the Southern Urals, and in the oak forests (F2), growing on the western slope but in the forest-steppe sites.

The use of the Jacquard index (**Table 3**) determining the proportion out of the total species list for two forest types which is common to both revealed low

Forest type	F1	F2	F3	F4	F5	F6	F7	F8	F9	F10	F11
Number of species	74	65	71	81	121	124	69	91	82	64	67
% of liverworts	13	14	7	10	19	17	22	14	9	8	15
% of mosses	87	86	93	90	81	83	78	86	92	92	85
Marchantiophyta											
Number of species	10	9	5	8	23	21	15	13	7	5	10
Number of genera	8	8	4	7	18	15	12	9	5	4	10
Number of families	7	6	4	6	14	10	8	7	5	4	8
Bryophyta											
Number of species	64	56	66	73	98	103	54	78	75	59	57
Number of genera	47	40	42	55	67	68	36	55	53	40	42
Number of families	23	22	25	28	32	34	20	29	27	21	22
% of pleurocarpous	66	59	61	59	54	49	46	45	55	53	60
% of acrocarpous	34	41	39	41	46	52	56	55	45	48	40

Table 2.
Some bryophyte richness indicators of the forest types studied (F1–F11) in the Southern Ural Mountains, Russia.

	F1	F2	F3	F4	F5	F6	F7	F8	F9	F10	F11
F1		0.34	0.37	0.45	0.39	0.36	0.25	0.33	0.26	0.33	0.37
F2	0.34		0.46	0.46	0.41	0.31	0.30	0.44	0.47	0.48	0.47
F3	0.37	0.46		0.46	0.39	0.34	0.28	0.38	0.39	0.44	0.37
F4	0.45	0.46	0.46		0.52	0.49	0.35	0.48	0.48	0.51	0.48
F5	0.39	0.41	0.39	0.52		0.64	0.38	0.49	0.44	0.40	0.45
F6	0.36	0.31	0.34	0.49	0.64		0.40	0.46	0.39	0.39	0.38
F7	0.25	0.30	0.28	0.35	0.38	0.40		0.40	0.30	0.33	0.36
F8	0.33	0.44	0.38	0.48	0.49	0.46	0.40		0.49	0.55	0.49
F9	0.26	0.47	0.39	0.48	0.44	0.39	0.30	0.49		0.51	0.48
F10	0.33	0.48	0.44	0.51	0.40	0.39	0.33	0.55	0.51		0.51
F11	0.37	0.47	0.37	0.48	0.45	0.38	0.36	0.49	0.48	0.51	

Table 3.
Jaccard index (similarity matrix) of bryophytes for different forest types in the Southern Ural Mountains, Russia.

similarity of bryophytes of different forests: the values of the index varied from 0.25 to 0.64. Species compositions of spruce-fir and dark coniferous broad-leaved forests (F5 and F6) are more similar as well as of various pine forests (F8–F10).

These relatively low values of Jaccard index can be explained by the high proportion of species with low constancy (**Table 4**), because, in general, the bryophytes have the scattered distribution within landscapes and vegetation types [22]. In the investigated forests, about 25% of species were revealed from one to three times. The proportion of species that were found less in 10 sample plots was higher in the xero-mesophytic oak, pine, larch, and birch forests growing on the border of forest and steppe (F2, 66%; F10, 66%; F9, 60%). Species with high frequency are few and consist of 5–15% of the bryophyte flora of different forest types.

Bryophytes are well adapted to the forest habitats and can be classified according to the substrate they live on as ground or epigeic (growing on the soil), epiphytic (on the bark of living trees), epixylic (on deadwood), or epilithic (on rock surfaces). Many bryophyte species are able to live on different substrates [23, 24]. It was shown that tree bases are a transition zone between the species growing on the rotten wood (two-thirds of epixylic bryophytes were found there) and epiphytic species growing on upper part of trunks. Also, almost one-third of species of the forest floor were revealed on decaying wood [11].

In all studied forests, xero-mesophytic species *Hypnum pallescens* (Hedw.) P. Beauv. *Sanionia uncinata* (Hedw.) Loeske, *Dicranum montanum* Hedw., *Sciurohypnum reflexum* (Starke) Ignatov & Huttunen, *Ptilidium pulcherrimum* (Weber.) Vain., *Brachythecium salebrosum* (Hoffm. ex F. Weber & D. Mohr) Schimp., *Callicladium haldanianum* (Grev.) H. A. Crum, and some others were found mainly on tree bases and the logs and stumps of initial stages of decaying. These species often grow on the bark of birch and are able to survive in relatively xeric environmental conditions of forest-steppe. The group of epixylic bryophytes growing on rotten wood of the last stages of destruction is not large. These species (*Blepharostoma trichophyllum* (L.) Dumort., *Lepidozia reptans* (L.) Dumort., *Lophoziopsis longidens* (Lindb.) Konstant. et Vilnet, *Lejeunea cavifolia* (Ehrh.) Lindb., *Dicranum flagellare* Hedw., *Tetraphis pellucida* Hedw., etc.) are typical mainly for old-growth spruce-fir and mixed forests (F5 and F6), where the proportion of epixylic species is higher (as compared with other forest types).

Forest type	F1	F2	F3	F4	F5	F6	F7	F8	F9	F10	F11
Number of relevés	96	115	287	139	195	132	52	120	88	101	179
Number of bryophytes	74	65	71	81	121	124	69	91	82	64	67
Epiphytes											
Pylaisia polyantha (Hedw.) Schimp.	III	III	III	II	II	I	+	I	I	III	II
Pseudoleskeella nervosa (Brid.) Nyholm	II	III	III	II	II	I	.	+	I	I	I
Orthotrichum speciosum Nees	I	+	I	I	I	.	.	r	r	r	r
Orthotrichum obtusifolium Brid.	r	r	I	r	r	I	.	.	.	.	.
Neckera pennata Hedw.	I	r	I	I	I	.	.	.	.	.	.
Homalia trichomanoides (Hedw.) Brid.	I	.	I	r	I	I	.	.	.	.	.
Leucodon sciuroides (Hedw.) Schwaegr.	.	I	III	I	I	.	.	.	I	.	.
Anomodon longifolius (Schleich. ex Brid.) Hartm.	.	r	I	I	I	I	.	.	+	+	.
Anomodon viticulosus (Hedw.) Hook. & Taylor	.	+	I	.	I	.	.	r	+	r	.
Serpoleskea subtilis (Hedw.) Loeske	.	r	I	I	+	+	I	r	.	+	.
Dicranum viride (Sull. & Lesq.) Lindb.	.	.	I	I	I	+	.	I	r	r	.
Frullania bolanderi Austin	.	r	.	r	I	+	r	.	.	.	I
Anomodon attenuatus (Hedw.) Huebener	.	.	r	r	.	I	.	.	.	.	.
Leskea polycarpa Hedw.	II	.	r	r	.	.	.	.	.	.	.
Haplocladium microphyllum (Hedw.) Broth.	.	.	r	r	.	.	.	.	r	.	r
Species occurring on the bases of trunks and rotten wood											
Hypnum pallescens (Hedw.) P. Beauv.	III	III	III	IV	V	II	III	III	V	IV	III
Ptilidium pulcherrimum (Weber.) Vain.	I	r	I	II	III	IV	IV	III	II	I	III
Lophocolea heterophylla (Schrad.) Dumort.	I	r	I	II	IV	II	II	I	r	+	II
Sanionia uncinata (Hedw.) Loeske	III	r	I	II	IV	IV	III	III	III	II	III
Sciuro-hypnum reflexum (Starke) Ignatov & Huttunen	III	II	IV	IV	IV	III	III	+	II	II	III
Dicranum montanum Hedw.	I	I	II	IV	IV	V	IV	IV	III	II	III
Brachythecium salebrosum (Hoffm. ex F. Weber & D. Mohr) Schimp.	III	II	III	III	III	I	r	II	II	III	II
Callicladium haldanianum (Grev.) H. A. Crum	II	r	II	IV	III	II	+	I	I	I	II
Amblystegium serpens (Hedw.) Schimp.	III	I	II	III	II	I	r	I	r	II	+
Platygyrium repens (Brid.) Schimp.	II	I	II	II	II	II	.	I	I	+	+
Campylidium sommerfeltii (Myrin) Ochyra	II	r	I	I	I	I	.	I	I	r	I
Plagiothecium laetum Schimp.	I	r	I	r	II	II	II	I	I	r	I
Plagiothecium denticulatum (Hedw.) Schimp.	I	r	I	r	II	II	I	I	r	.	.
Lophocolea minor Nees	I	r	II	I	II	I	+	+	+	r	r
Pohlia nutans (Hedw.) Lindb.	r	r	I	I	I	II	I	III	II	II	II
Brachytheciastrum velutinum (Hedw.) Ignatov & Huttunen	I	II	I	I	II	I	I	I	I	I	+

Forest type	F1	F2	F3	F4	F5	F6	F7	F8	F9	F10	F11
Radula complanata (L.) Dumort.	r	I	II	I	II	I	r	I	+	r	I
Sciuro-hypnum starkei (Brid.) Ignatov & Huttunen	.	.	I	r	I	II	+	r	r	r	I
Tetraphis pellucida Hedw.	r	.	r	r	I	II	I	I	.	.	r
Lepidozia reptans (L.) Dumort	.	r	.	.	+	II	r	.	.	.	r
Dicranum fragilifolium Lindb.	.	.	I	.	r	II	.	.	I	.	.
Blepharostoma trichophyllum (L.) Dumort	.	.	.	.	I	II	II	.	.	.	.
Lophoziopsis longidens (Lindb.) Konstant. & Vilnet	.	.	.	.	I	I	II	r	.	.	r
Bryum capillare Hedw.	I	r	r	r	.	.	.	r	r	r	.
Lophozia ventricosa (Dicks.) Dumort.	r	r		r	I	I	I	r	.	r	r
Bryum moravicum Podp.	.	r	I	r	I	I	I	+	+	r	.
Dicranum flagellare Hedw.	.	.	I	I	I	I	.	II	I	+	II
Cynodontium strumiferum (Hedw.) Lindb.	.	.	.	r	I	I	I	I	.	.	.
Orthocaulis attenuatus (Nees) A. Evans	.	.	.	.	+	I	I	r	.	.	.
Cephaloziella hampeana (Nees) Schiffn. ex Loeske	.	.	.	.	r	.	I	I	r	.	I
Eurhynchiastrum pulchellum (Hedw.) Ignatov & Huttunen	r	r	.	I	II	I	.	I	+	.	I
Epilithic species											
Paraleucobryum longifolium (Hedw.) Loeske	.	I	I	r	I	II	III	I	r	r	+
Tortella tortuosa (Hedw.) Limpr.	.	II	.	r	I	I	.	r	II	I	I
Schistidium apocarpum s.l. (Hedw.) Bruch & Schimp.	.	+	r	.	I	II	.	r	II	r	r
Homomallium incurvatum (Schrad. ex Brid.) Loeske	.	r	r	r	I	I	.	.	+	+	r
Hedwigia ciliata (Hedw.) P. Beauv.	.	I	.	r	.	I	.	I	+	r	r
Campyliadelphus chrysophyllus (Brid.) R. S. Chopra	.	.	.	r	I	II	.	II	r	r	.
Hypnum cupressiforme Hedw.	.	.	r	r	I	I	r	r	I	.	.
Pohlia cruda (Hedw.) Lindb.	.	.	.	.	I	II	.	I	r	r	r
Ditrichum flexicaule (Schwaegr.) Hampe	.	.	.	r	r	+	.	I	II	.	.
Oxystegus tenuirostris (Hook. & Taylor) A. J. E. Sm.	r	r	r	.	I	.	.	.	.	.	.
Sciuro-hypnum populeum (Hedw.) Ignatov & Huttunen	r	I	r	.	.	.	I	r	.	.	.
Pseudoleskeella tectorum (Funck ex Brid.) Kindb. ex Broth.	.	+	.	.	r	.	.	.	r	.	r
Distichium capillaceum (Hedw.) Bruch & Schimp.	.	.	.	r	I	I	.	.	+	.	.
Entodon schleicheri (Schimp.) Demet.	.	.	.	r	r	r	.	.	r	.	.
Bryoerythrophyllum recurvirostrum (Hedw.) P. C. Chen	.	.	.	.	r	+	.	r	r	.	.
Encalypta procera Bruch	.	.	.	.	r	r	.	.	+	.	.

Forest type	F1	F2	F3	F4	F5	F6	F7	F8	F9	F10	F11
Dicranum spadiceum J. E. Zetterst.	.	.	.	.	.	I	.	I	r	.	.
Dicranum flexicaule Brid.	.	.	.	r	.	r	r	.	r	r	.
Hypnum recurvatum (Lindb. & Arnell) Kindb.	r	r	.	.	r	r	r	.	r	.	r
Mnium stellare Hedw	.	.	.	r	+	I	.	+	.	.	.
Epigeic species sometimes found on stones and fallen trees											
Brachythecium mildeanum (Schimp.) Schimp.	II	.	.	r	I	r	.	.	r	.	.
Brachythecium rivulare Schimp.	II	.	I	.	.	r	.	.	.	.	.
Calliergonella lindbergii (Mitt.) Hedenaes	II	.	.	r	r	+	.	.	.	.	.
Plagiomnium ellipticum (Brid.) T. J. Kop.	II	.	r	r	+	r	.	r	.	r	.
Calliergon cordifolium (Hedw.) Kindb.	II	.	.	.	r	r	.	.	.	.	.
Campylium stellatum (Hedw.) Lange & C. E. O. Jensen	r	.	.	.	r	r	.	.	r	.	r
Rhizomnium pseudopunctatum (Bruch & Schimp.) T. J. Kop.	r	.	.	r	I	+	r	.	.	.	.
Fissidens taxifolius Hedw.	I	.	II	r	I	r	.	r	.	.	I
Oxyrrhynchium hians (Hedw.) Loeske	I	.	I	.	II	I	.	r	r	r	.
Plagiomnium cuspidatum (Hedw.) T. J. Kop.	III	II	IV	II	IV	II	+	I	+	I	II
Pleurozium schreberi (Willd. ex Brid.) Mitt.	I	r	r	II	IV	V	V	V	V	IV	V
Ptilium crista-castrensis (Hedw.) De Not.	r	.	.	II	II	IV	V	III	III	I	II
Dicranum polysetum Sw. ex anon.	r	r	.	II	I	III	II	IV	IV	II	II
Dicranum scoparium Hedw.	r	I	I	II	III	IV	V	IV	IV	II	III
Hylocomium splendens (Hedw.) Schimp.	.	.	.	II	III	V	V	IV	III	II	II
Rhytidiadelphus triquetrus (Hedw.) Warnst.	.	.	r	II	III	III	r	III	II	I	III
Plagiochila porelloides (Torr. ex Nees) Lindenb.	I	.	.	r	I	III	.	.	.	.	.
Sciuro-hypnum curtum (Lindb.) Ignatov	I	r	I	I	III	II	I	II	I	I	III
Rhodobryum roseum (Hedw.) Limpr.	I	I	.	r	II	II	.	I	.	+	III
Hylocomiastrum pyrenaicum (Spruce) M. Fleisch.	.	r	r	.	II	I	I	r	.	.	.
Hylocomiastrum umbratum (Hedw.) M. Fleisch.	.	.	.	r	II	II	II	.	.	.	.
Cirriphyllum piliferum (Hedw.) Grout	I	r	.	r	II	I	.	r	.	.	r
Thuidium assimile (Mitt.) A. Jaeger	r	.	.	.	I	II	.	I	.	r	.
Climacium dendroides (Hedw.) F. Weber & D. Mohr	I	.	r	r	I	r	.	+	.	+	r
Atrichum undulatum (Hedw.) P. Beauv.	r	.	r	.	I	I	+	r	r	.	.
Plagiomnium rostratum (Schrad.) T. J. Kop.	I	.	r	r	+	.	.	.	.	.	r

Forest type	F1	F2	F3	F4	F5	F6	F7	F8	F9	F10	F11
Plagiomnium medium (Bruch & Schimp.) T. J. Kop.	r	.	.	.	I	I	.	r	.	.	I
Brachythecium rutabulum (Hedw.) Schimp.	.	r	r	.	I	r	.	.	.	r	.
Polytrichum juniperinum Hedw.	.	+	.	r	+	I	I	II	+	+	+
Dicranum fuscescens Sm.	.	.	.	r	I	I	I	I	II	+	I
Rhytidiadelphus subpinnatus (Lindb.) T. J. Kop.	r	.	.	.	I	II	.	r	.	.	.
Barbilophozia lycopodioides (Wallr.) Loeske	.	.	.	.	I	I	II	.	.	.	.
Barbilophozia hatcheri (A. Evans) Loeske	.	.	.	.	I	I	II	.	.	.	.
Barbilophozia barbata (Schmidel ex Schreb.) Loeske	.	I	.	r	I	I	.	I	+	.	.
Polytrichastrum longisetum (Sw. ex Brid.) G. L. Sm.	.	.	.	.	I	I	I	.	.	.	.
Dicranum majus Sm.	.	.	r	.	.	I	I	.	.	.	.
Polytrichastrum densifolium Wilson ex Mitt.	.	.	.	.	I	.	I	.	.	.	.
Atrichum flavisetum Mitt.	r	.	.	r	I	I	.	.	.	.	r
Abietinella abietina (Hedw.) M. Fleisch.	.	I	.	r	I	I	.	I	II	+	r
Rhytidium rugosum (Hedw.) Kindb.	.	r	.	r	.	.	.	III	I	.	.
Polytrichum piliferum Hedw.	.	.	.	.	.	.	r	II	.	r	.
Ceratodon purpureus (Hedw.) Brid.	.	+	r	r	I	.	.	I	II	II	I
Brachythecium albicans (Hedw.) Schimp.	.	I	I	.	r	.	.	I	r	+	r
Syntrichia ruralis (Hedw.) F. Weber & D. Mohr	.	I	.	.	.	.	.	.	I	+	.
Leptobryum pyriforme (Hedw.) Wilson	.	.	.	.	.	r	r	.	r	.	r
Dicranella heteromalla (Hedw.) Schimp.	.	.	.	.	r	.	r	I	.	.	r

Note: The species with low constancy are excluded.

Table 4.
Shortened synoptic table of bryophytes revealed in forest communities in the Southern Ural Mountains, Russia.

The "real" epiphytes, growing on the bark of the living trees at a height of 1 m and above, are *Pylaisia polyantha* (Hedw.) Schimp., *Pseudoleskeella nervosa* (Brid.) Nyholm, *Orthotrichum obtusifolium* Brid., and *O. speciosum* Nees. These species are xero-mesophytic and often may be found on the trees of settlements and in contour strip forests surrounded by agricultural lands. There is another group of mesophytic epiphytes growing mainly on the bark of old broad-leaved trees. These species (*Leucodon sciuroides* (Hedw.) Schwaegr., *Neckera pennata* Hedw., *Homalia trichomanoides* (Hedw.) Brid., *Anomodon longifolius* (Schleich. ex Brid.) Hartm., *Dicranum viride* (Sull. & Lesq.) Lindb., and some others) are quite sensitive to the fluctuations in temperature and humidity of habitats. An analysis of the distribution of these species using ecological scales showed that they have a narrow ecological amplitude, especially in relation to the factor of continentality [25, 26]. Many of these species are basophilous and can grow on limestone outcrops. Probably, the

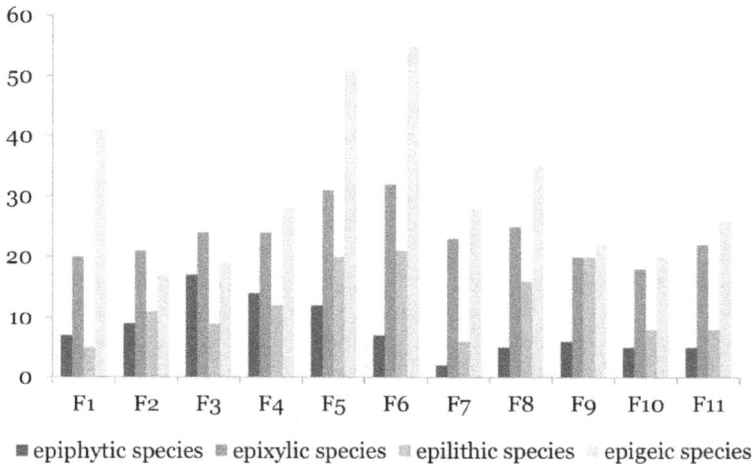

Figure 2.
Substrate groups of bryophytes in the different forest types of the Southern Ural Mountains, Russia.

presence of limestones helps preserve the local subpopulations of these species in the felling areas. The proportion of epiphytic species in the bryophyte flora of different forest types is more significant in broad-leaved and mixed pine broad-leaved forests (F3 and F4) (**Figure 2**).

Projective cover of epigeic bryophytes varies greatly depending on the forest type, being insignificant in broad-leaved forests (<1%) and reaching the higher values in boreal spruce-fir and pine forests (up to 80%) (**Table 1**). The high species richness of epigeic bryophytes in tall-herb dark coniferous and mixed forests (F5 and F6), as well as in floodplain alder forests (F1), may be explained by high humidity of soil in these habitats. In relatively dry habitats of thermophytic oak forests (F2) with dense herb layer and in shaded mesic linden-maple-oak forests (F3), the diversity of epigeic species is quite low (**Figure 2**).

Boreal species *Pleurozium schreberi* (Willd. ex Brid.) Mitt., *Hylocomium splendens* (Hedw.) Schimp., *Dicranum polysetum* Sw. ex anon., *Dicranum scoparium* Hedw., and *Rhytidiadelphus triquetrus* (Hedw.) Warnst. are almost completely absent in the oak forests (F2), occasionally occur in floodplain forests (F1), and are quite common in other forest types. In "typical" boreal green-moss coniferous forests (F7 and F8), they grow mostly on the forest floor, but in other forest types, these species were found mainly on logs, bases of trees, and stones where they avoid the competition with vascular plants.

In many forests of the Southern Urals, there are numerous rock outcrops. The proportion of epilithic species is high in the hemiboreal xero-mesophytic pine and pine-larch forests (F9) growing on steep southern and southeastern slopes with lime outcrops, where these species consist of 33% of bryophyte flora, but total species richness of epilithic bryophytes is highest in the dark coniferous forests (**Figure 2**).

5. The peculiarities of bryophyte composition in the secondary forests

Epiphytic, epixylic, and epilithic bryophytes are particularly responsive to microclimatic as well as physical and chemical substrate properties, which directly depend on tree age and diameter, bark texture, or decay stages of deadwood. Many bryophytes are sensitive to forest management practices [27, 28], but in Russia, the influence of felling on these components of forest vegetation has not been well studied [8].

A multitude of restoration practices are currently being used in the Southern Ural forests, but in spite of the government recommendations to establish in the felling areas the forest plantations through human-induced direct planting or seeding, the large felled areas are overgrown by deciduous pioneer tree species, for example, birch and aspen due to natural regeneration. Understanding how secondary forests differ from the indigenous forests in terms of diversity, structure, and function provides the basis for forest restoration ecology. The availability of mountain boreal forests in different stages of succession provides an opportunity for comparing of plant diversity and the structural and functional elements in the Ural's indigenous forests as well as in the secondary forests established after clear-cutting. Such data are crucial for planning monitoring and restoration of unique mountain forests [29].

In the Southern Urals, the study of reforestation processes has been only recently begun, but some preliminary results concerning dynamic of bryophyte diversity during natural forest regeneration were obtained [11, 26, 30].

The study of secondary plant communities originated after clear-cutting in the pine and broad-leaved forests [26, 30] has shown that bryophytes of various substrate groups respond to felling in different ways. Xero-mesophytic epiphytic and epixylic species (*Pylaisia polyantha* (Hedw.) Schimp., *Pseudoleskeella nervosa* (Brid.) Nyholm, *Brachythecium salebrosum* (Hoffm. ex F. Weber & D. Mohr) Schimp., *Amblystegium serpens* (Hedw.) Schimp., *Sciuro-hypnum reflexum* (Starke) Ignatov & Huttunen, and *Hypnum pallescens* (Hedw.) P. Beauv., etc.) seem to be more tolerant to habitat changes after felling. Usually, these species sharply reduce their constancy after felling but begin to grow actively after 3–4 years on stumps and felling residues and on the bases of young tree trunks. In secondary forests they restore or even increase their constancy.

The light-demanding colonists (*Ceratodon purpureus* (Hedw.) Brid., *Bryum caespiticium* Hedw., *Funaria hygrometrica* Hedw., *Pogonatum urnigerum* (Hedw.) P. Beauv., *Dicranella heteromalla* (Hedw.) Schimp, *Leptobryum pyriforme* (Hedw.) Wilson, and some others) have a relatively high constancy during the first 7–20 years after felling, but later they belong to the category of sporadically occurring species, growing on the soil near the roots of fallen trees and forest roadsides.

The nemoral species *Orthotrichum speciosum* Nees., *Leucodon sciuroides* (Hedw.) Schwaegr., *Dicranum viride* (Sull. & Lesq.) Lindb., *Neckera pennata* Hedw., *Homalia trichomanoides* (Hedw.) Brid., *Anomodon longifolius* (Schleich. ex Brid.) Hartm., *Oxyrrhynchium hians* (Hedw.) Loeske, and *Fissidens taxifolius* Hedw. are vulnerable to tree felling. They either disappear or have little constancy in the early succession communities.

Boreal species *Pleurozium schreberi* (Willd. ex Brid.) Mitt. and *Hylocomium splendens* (Hedw.) Schimp. are more resistant against changes in ecological regimes of habitats. *Ptilium crista-castrensis* (Hedw.) De Not., *Dicranum polysetum* Sw. ex anon., and *Rhytidiadelphus triquetrus* (Hedw.) Warnst. are more vulnerable [26].

In general, changing of bryophyte diversity in the felled areas seems to be as follows:

During 1–4 years after felling, the species richness is sharply reduced, in some cases by 50–70%. The cover of herb layer increases due to disturbances of the forest floor and high illumination. These processes are particularly intense after summer clear-cutting connected with strong disturbances of the forest floor. Usually, species richness of bryophytes begins to rise only after 7 years or more, when young trees appear and shading becomes more or less significant.

Nearly 60–90 years after felling, the similarity of the bryophyte composition in the indigenous and mature secondary forests still remains quite low due to the differences in species composition. Considering the fact that a significant number of bryophytes have a low abundance and scattered distribution, it is often difficult to say whether the disappearance of particular species is due to the felling or for some other reason.

6. Rare bryophytes in the forests of the Southern Urals

As mentioned above, about 25% of bryophyte species revealed in the sample plots within indigenous forests of the Southern Urals were found one to three times and formally may be considered as "rare" species. This group comprises *Riccia rhenana* Lorb. ex Muell. Frib., *Racomitrium aciculare* (Hedw.) Brid., *R. aquaticum* (Brid. ex Schrad.) Brid., *Odontoschisma denudatum* (Mart.) Dumort., *Tetraplodon angustatus* (Hedw.) Bruch & Schimp., *Philonotis seriata* Mitt., *Myrinia pulvinata* (Wahlenb.) Schimp., *Jungermannia pumila* With., *Rhizomnium magnifolium* (Horik.) T. J. Kop., *Brachythecium laetum* (Brid.) Schimp., and many other species, some of whom are not typical for the forests. Also, it should be noted that the small number of their known localities may be explained with insufficient botanical knowledge of the region. In this connection, the selection of indicator species for identifying forest areas that need protection constitutes the important challenge.

There are two aspects—the identification of indicator species of old-growth and primeval forests. Often, these two concepts are confused, which is not quite justified due to the differences in the life strategies of bryophytes and duration of the undisturbed existence of forests. The identification of areas where ecological conditions of habitats (especially humidity and illumination) were not changed for a very long time is very important for the conservation of bryophytes. Such areas are called zones of ecological continuity [31]. Not all old-growth forests are suitable for this purpose. Usually the term "old-growth forest" is defined as the forest in which the age of tree stand is more than 120–260 years (depending on the dominant tree species and the region) [8]. At the same time, the survival of relic populations of bryophytes having a limited dispersal activity and high demands on the stability of ecological regime in the habitats is possible mostly in primeval or ancient forests.

In authors' opinion, under selection of such criteria, both reproduction and the reasons of rarity of species should be taken into account. Among the bryophytes considered as indicators of old-growth forests in North-West European Russia [8], there are the species with different reproduction characteristics and dispersal ability. For instance, according to the system of life strategy [32], such species as *Homalia trichomanoides* (Hedw.) Brid., *Orthotrichum affine* Schrad. ex Brid., and *Haplocladium microphyllum* (Hedw.) Broth. may be considered as colonists and *Leucodon sciuroides* (Hedw.) Schwaegr., *Frullania bolanderi* Austin, and *Lejeunea cavifolia* (Ehrh.) Lindb. as shuttles. There are some species that may be considered as perennial stayers because of their low sporophyte frequency or lacking of sporophyte in study area but have vegetative propagules (*Dicranum viride* (Sull. & Lesq.) Lindb., *Barbilophozia attenuata* (Mart.) Loeske, etc.).

Most of these species are epiphytic or epixylic and can spread to a new site from the nearby forest area, if the suitable substrates are available. These species may be considered as indicators of old-growth forests. On the other hand, stayer species that have no vegetative propagules and prefer to grow on the soil or stones could be seen as indicators of both old-growth and primeval forests. This group includes the species that are very rare in the Southern Urals (*Eurhynchium angustirete* (Broth.) T. J. Kop., *Dicranum drummondii* Muell. Hal., *Entodon schleicheri* (Schimp.) Demet., *Plagiomnium confertidens* (Lindb. & Arnell.) T. J. Kop., etc.). Some of them may be considered as relicts.

Species	Reproduction features	Reasons of rarity
Indicators of old-growth forests		
Neckera pennata Hedw.	1	2, 4
Homalia trichomanoides (Hedw.) Brid.	1	2, 4
Haplocladium microphyllum (Hedw.) Broth.	1	2, 3
Dicranum viride (Sull. & Lesq.) Lindb.	3	1, 2, 4
Anomodon longifolius (Schleich. ex Brid.) Hartm.	4	2, 4
Lejeunea cavifolia (Ehrh.) Lindb.	3	1, 3
Calypogeia integristipula Steph.	3	3, 4
Orthocaulis attenuatus (Nees) A. Evans	3	3, 4
Lepidozia reptans (L.) Dumort.	4	3, 4
Polytrichastrum pallidisetum (Funck) G. L. Sm.	4	4
Eurhynchium angustirete (Broth.) T. J. Kop.	4	4
Dicranum drummondii Muell. Hal.	4	1
Hylocomiastrum umbratum (Hedw.) M. Fleisch.	4	1
Iwatsukiella leucotricha (Mitt.) W. R. Buck & H. A. Crum	4	1
Rare species with disjunctive area of distribution		
Entodon schleicheri (Schimp.) Demet.	1	1, 4
Brachythecium geheebii Milde	4	1
Anomodon rugelii (Muell. Hal.) Keissl.	4	1, 4
Orthothecium intricatum (Hartm.) Schimp.	4	1
Myurella sibirica (Muell. Hal.) Reimers	4	1
Plagiomnium confertidens (Lindb. & Arnell) T. J. Kop.	4	1
Campylidium calcareum (Crundw. & Nyholm) Hedenaes	4	1, 4

Note: *Reproduction features: 1, sporophytes are frequent, and vegetative propagules are absent; 2, sporophytes are frequent, and vegetative propagules are present; 3, sporophytes are rare or unknown in study area, and vegetative propagules are present; 4, sporophytes are rare or unknown in study area, and vegetative propagules are absent. Reasons of rarity: 1, rare species growing at the border of the range (or with a disjunctive area); 2, species preferring to grow on old broad-leaved trees; 3, species growing on decaying wood of last stages of destruction; 4, sciophytic and hygrophilous species disappearing during the felling and clearing of the forests.*

Table 5.
Hemerobic bryophytes in the forests of the Southern Ural Mountains, Russia.

In **Table 5**, some species that may be considered as hemerobic in the Southern Urals are shown. It should be noted that the indicator value of these taxa is regional and may be different in the other areas where climatic conditions are not such continental as in the Southern Urals.

The forest associations characterized by high diversity and concentration of rare bryophytes were identified. There are mainly tall-herb and mixed forests (F5 and F6) where regionally rare species (*Brachythecium geheebii* Milde, *Eurhynchium angustirete* (Broth.) T. J. Kop., *Plagiomnium confertidens* (Lindb. & Arnell.) T. J. Kop., *Haplocladium microphyllum* (Hedw.) Broth., *Polytrichastrum pallidisetum* (Funck) G. L. Sm., *Iwatsukiella leucotricha* (Mitt.) W. R. Buck & H. A. Crum, and some others) were found. The proportion of rare species in these forests is about of 9%. In contrary, the bryophyte flora of pine forests seems to be well adapted to regular disturbances, that is, fires, and contain a significant number of species with high reproduction activity. Also, these forests are characterized by a low number of rare species that consists of about 1% of bryophyte flora.

In the Pleistocene, refugia of the nemoral flora of the western slope of Southern Urals contained both the nemoral and the black taiga floristic elements [33]. Our previous research had shown some similarities between bryophyte flora of tall-herb broad-leaved dark coniferous forests of the Southern Urals and the black taiga of Salair Ridge in Siberia [11], as well as the availability of a significant number of relicts both of European and Asian origins in these forests. The sites of these forests are most valuable for nature conservation and should be protected.

Acknowledgements

The authors are grateful to the Russian Foundation for Basic Research for supporting of this study (grants no. 18-04-00641 and 16-04-00985-a) and governmental contract #075–00326–19–00 theme AAAA–A118022190060–6.

Appendices and nomenclature

The species names are given in accordance with the checklists for the territory of Russia [34, 35] and some later sources [36].

Author details

Elvira Baisheva*, Pavel Shirokikh and Vasiliy Martynenko
Institute of Biology—Subdivision of the Ufa Federal Research Centre of the Russian Academy of Sciences, Ufa, Russia

*Address all correspondence to: elvbai@mail.ru

IntechOpen

References

[1] Krasheninnikov I, Kucherovskaya-Rozhanets S. Prirodnye resursy Bashkirskoy ASSR [Natural Resources of the Bashkir ASSR]. Izdatel'stvo AN SSSR: Moscow-Leningrad; 1941. p. 1, 156

[2] Yaparov I, editor. Atlas Respubliki Bashkortostan [Atlas of the Republic of Bashkortostan]. GUP GRI "Bashkortostan": Ufa; 2005. p. 420

[3] Gorchakovskiy P. Rastitel'nost' i botaniko-geograficheskoye deleniye Bashkirskoy ASSR [Vegetation and botanical-geographical division of the Bashkir ASSR]. In: Alekseyev YU, editor. Opredelitel' vyshikh rasteniy Bashkirskoy ASSR [The Handbook of Higher Plants of the Bashkir ASSR]. Moscow: Izdatel'stvo Nauka; 1988. pp. 5-13

[4] Mirkin B, editor. Flora i rastitel'nost' Yuzhno-Ural'skogo gosudarstvennogo prirodnogo zapovednika [Flora and Vegetation of the South Ural State Nature Reserve]. Ufa: Izdatel'stvo Gilem; 2008. p. 528

[5] Mirkin B, editor. Flora i rastitel'nost' Natsional'nogo parka «Bashkiriya» (sintaksonomiya, antropogennaya dinamika, ekologicheskoye zonirovaniye) [Flora and Vegetation of the National Park "Bashkiria" (Syntaxonomy, Anthropogenic Dynamics, Ecological Zoning)]. Ufa: Gilem; 2010. p. 512

[6] Halme P, Allen KA, Aunins A, Bradshaw RHW, Brumelis G, Cada V, et al. Challenges of ecological restoration: Lessons from forests in northern Europe. Biological Conservation. 2013;**167**:248-256. DOI: 10.1016/j.biocon.2013.08.029

[7] Shirokikh P, Martynenko V, Kunafin A. Experience in syntaxonomic and ordination analysis of progressive succession in cutover areas of boreal light conifer forests in the Southern Urals. Russian Journal of Ecology. 2013;**34**:185-192. DOI: 10.1134/S1067413613030120

[8] Andersson L, Alekseeva NM, Kuznetsova ES. Vyyavlenie i obsledovanie biologicheski tsennykh lesov na Severo-Zapade yvropeiskoi chasti Rossii [Identification and Investigation of Biologically Valuable Forests in the North-East of European Russia]. Vol. 1. St. Petersburg: Pobeda; 2009. p. 238

[9] Yaroshenko A, Potapov PV, Turubanova SA. Malonarushennye lesnye territorii Evropeiskogo Severa Rossii [Scarce-Disturbed Forest Areas of the North of European Russia]. Moscow: Greenpeace Russia; 2001. p. 75

[10] Martynenko V. Sintaksonomiya lesov Yuzhnogo Urala kak teoreticheskaya osnova razvitiya sistemy ikh okhrany [Syntaxonomy of forests of the Southern Urals as a theoretical basis for the development of their protection] [thesis]. Ufa: Bashkir State University; 2009

[11] Baisheva EZ, Martynenko VB, Shirokikh PS. Malonarushennye lesnye territorii Evropeiskogo Severa Rossii [Scarce-Disturbed Forest Areas of the North of European Russia]. Ufa: Gilem; 2015. p. 352

[12] Kraus D, Krumm F, editors. Integrative Approaches as an Opportunity for the Conservation of Forest Biodiversity. Freiburg: European Forest Institute; 2013. p. 284

[13] Maksimov AI, Potemkin AD, Hokkanen TJ, Maksimova T. Bryophytes of fragmented old-growth spruce forest stands of the North Karelian Biosphere Reserve and adjacent areas of Finland. Arctoa. 2003;**12**:9-23

[14] Degteva SV, Zheleznova GV, Pystina TN, Shubina TP. Tsenoticheskaya i floristicheskaya structura listvennykh lesov Evropeiskogo Severa [Coenotic and Floristic Structure of Deciduous Forests of European North]. St. Petersburg: Nauka; 2001. p. 269

[15] Smirnova OV, editor. Vostochnoevropeiskie lesa: istoriya v golotsene i sovremennost' [Eastern European Forests: Holocene History and Present Status]. Vol. 1. Moscow: Nauka; 2004. p. 479

[16] Muldashev A, editor. Reestr osobo okhranyaemykh territoriy Respubliki Bashkortostan [Registry of Nature Protected Areas of Republic of Bashkortostan]. 2nd ed. Ufa: MediaPrint; 2010. p. 414

[17] Braun-Blanquet J. Pflanzensoziologie. Grundzuge der Vegetationskunde. Wien-New York: Springer-Verlag; 1964. p. 865

[18] Mucina L, Bültmann H, Dierßen K, Theurillat JP, Raus T, Carni A, et al. Vegetation of Europe: Hierarchical floristic classification system of vascular plant, bryophyte, lichen, and algal communities. Applied Vegetation Science. 2016;19:1-264. DOI: 10.1111/avsc.12257

[19] Hennekens SM, Scaminée JHJ. TURBOVEG, a comprehensive data base management system for vegetation data. Journal of Vegetation Science. 2001;12:589-591. DOI: 10.2307/3237010

[20] Tichý L, Holt J, Nejezchlebová M. JUICE. Program for Management, Analysis and Classification of Ecological Data. 2nd ed. Brno: Masaryk Univ. Press; 2011. p. 65

[21] Zverev AA. Informatsionnye tekhnologii v issledovaniyakh rastitel'nogo pokrova [Information Technologies in the Investigations of Plant Cover]. Tomsk: TML-Press; 2007. p. 304

[22] Vitt DH, Halsey LA, Bray J, Kinser A. Patterns of bryophyte richness in a complex boreal landscape: Identifying key habitats at McClelland Lake Wetland. The Bryologist. 2003;106:372-382. DOI: 10.1639/03

[23] Barkman J. Phytosociology and Ecology of Cryptogamic Epiphytes. Assen: Van Gorcum; 1958. p. 628

[24] Goffnet B, Shaw AJ, editors. Bryophyte Biology. 2nd ed. New York: Cambridge University Press; 2009. p. 565

[25] Baisheva E, Mežaka A, Shirokikh P, Martynenko V. Ecology and distribution of *Dicranum viride* (Sull. & Lesq.) Lindb. in the Southern Ural Mts. Arctoa. 2013;22:41-50

[26] Baisheva EZ, Shirokikh PS, Martynenko VB, Mirkin BM. Influence of clear fellings on the bryophyte component of the broad-leaved forests of the Bashkir Cis-Ural Region. Russian Journal of Ecology. 2018;49:1-9

[27] Vellak K, Ingerpuu N. Management effects on bryophytes in Estonian forests. Biodiversity and Conservation. 2005;14:3255-3263. DOI: 10.1007/s10531-004-0445-1

[28] Fenton NJ, Frego KA. Bryophyte (moss and liverwort) conservation under remnant canopy in managed forests. Biological Conservation. 2005;122:417-430. DOI: 10.1016/j.biocon.2004.09.003

[29] Ivanova N. Differentiation of forest vegetation after clear-cuttings in the Ural Mountains. Modern Applied Science. 2014;8:195-203. DOI: 10.5539/mas.v8n6p195

[30] Baisheva EZ, Shirokikh PS, Martynenko VB. Effect of clear-cutting

on bryophytes in the pine forests of the South Urals. Arctoa. 2015;**24**:547-555

[31] Norden B, Appelqvist T. Conceptual problems of ecological continuity and its bioindicators. Biodiversity and Conservation. 2001;**10**:779-791. DOI: 10.1023/A:101667510

[32] During HJ. Life strategies of bryophytes: A preliminary review. Lindbergia. 1979;**5**:2-18

[33] Kamelin RV, Ovesnov SA, Shilova SI. Nemoral'nye relikty vo florakh Urala I Sibiri [Nemoral Relicts in the Floras of Ural and Siberia]. Perm': Perm' University; 1999. p. 82

[34] Ignatov MS, Afonina OM, Ignatova EA, et al. Check-list of mosses of East Europe and North Asia. Arctoa. 2006;**15**:1-130. DOI: 10.15298/ arctoa.15.01

[35] Konstantinova NA, Bakalin VA, et al. Checklist of liverworts (Marchantiophyta) of Russia. Arctoa. 2009;**18**:1-64. DOI: 10.15298/ arctoa.18.01

[36] Ignatov MS, editor. Moss flora of Russia. In: Oedipodiales—Grimmiales. Vol. 2. Moscow: KMK Scientific Press Ltd; 2017. p. 560

Species Distribution Patterns in Subgenus Cuspidata (Genus Sphagnum L.) on the East European Plain and Eastern Fennoscandia

Sergei Yu. Popov

Abstract

The geographic range of 13 species from the subgenus Cuspidata in the East European Plain and Eastern Fennoscandia has been studied. Model maps for each species occurrence were constructed using geostatistics techniques (kriging method). Continuous coverages of 23 climatic factors were used in analysis also. We used dataset that proposed by authors of program WORLDCLIM. To learn how corresponding values of climatic factors and species occurrence correlation and cluster analysis were conducted. It was found that 7 of 13 species are widespread on the East European Plain and Eastern Fennoscandia, and 6 species have the restricted ranges. Values of occurrence of all species (except *Sphagnum lenense*) have a strong correlation with moisture factors (relative air humidity and sum of precipitation) in summer-autumn period. Such preferences allow them to grow successfully in Subarctic and Baltic regions, where high climatic humidity is observed. Restricted species are concentrated around the Baltic Sea and zones of the highest occurrence of widespread species are located at the same region. All species can be divided into four clusters according to its climatic preferences. Distribution of such species as *S. obtusum* seems to be strongly associated with two tongues of the Last Glacier, and this species seems to be a glacial relic.

Keywords: Sphagnum, Cuspidata, biogeography, BIOCLIM, distributional range, GIS, geostatistics, kriging method

1. Introduction

Sphagnum mosses are widely distributed plants in wet habitats. They are edificators in boggy forests and bogs in all plant zones. The ecology of species of the genus Sphagnum is now well studied, and environmental factors that play a leading role in the division of ecological space among Sphagnum species are well known [1–11]. Until now, however, the question about the division of geographical space by species remains open, especially due to the influence of climatic factors. There are two principal works on the biogeography of the genus Sphagnum [12, 13], which consider the geographical variability of species diversity of the genus in Western

Europe by methods of zonal statistics, that is, within the administrative boundaries of administrative states. In both cited works, the authors find the center of species diversity of the genus Sphagnum in the Scandinavian Peninsula. There does not seem to be any work that considers the distribution of species within its natural boundaries. Therefore, the present article is intended to fill this gap for the territory of the East European plain and Eastern Fennoscandia. As more than 50 species of Sphagnum grow in Europe [14], it is not possible to consider all of them in a single article due to lack of space. Therefore, in this chapter, we consider the distribution of species of the subgenus Cuspidata only, growing on the territory of the East European plain and Eastern Fennoscandia (EEPEF). In Europe (from the Atlantic to Urals), there are 17 species of the subgenus Cuspidata [14]. Only 14 species occur in the EEPEF. These are as follows: *Sphagnum angustifolium, S. annulatum, S. balticum, S. cuspidatum, S. fallax, S. flexuosum, S. jensenii, S. lenense, S. lindbergii, S. majus, S. obtusum, S. pulchrum, S. riparium*, and *S. tenellum*. Although some species are difficult to identify, these errors are easy to identify and correct by comparing bulk materials from different geographic locations. Moreover, a mathematical method for modeling maps, which is used in this work—the kriging method [15–20] serves as error protection. This method is widely used to build maps of temperature distribution in climatology, compiling digital elevation models in geodesy, etc. The advantages of this modeling method, compared to other ones currently used, are discussed in detail in previously published paper [21]. In bryology and biogeography, we use the kriging method for the first time. In short, the kriging method allows us to create model maps of the species distribution, which can reflect not only the boundaries of the species range as a whole, but also the species activity within the range. In addition, taking into account the weights of input points, values allow to cut off the noise while maintaining the overall trend of the distribution of the species. In the case of the study of mosses distribution, random incorrect definitions of species in some geographic points just appear as noise on a mathematical surface. All of the above is true for such species for which we have data set from the entire study area. Among the 14 species of the subgenus Cuspidata which is found in European Russia and adjacent countries, only one species does not satisfy this condition. This is *Sphagnum annulatum*. Since the valid description of this species was made relatively recently [22], and actually in Russian local floras, it "appears" around the late of 1990s, the definitions of this species cannot cover the entire study area (the database of local floras includes works, which were conducted since 1960s till 2017). In this connection, in the present work, 13 species of Sphagnum of the subgenus Cuspidata (from the list above), excluding *S. annulatum*, are analyzed.

The purpose of the present work is to simulate the ranges of species and study their distribution patterns, in connection with spatial changes of climatic factors in the EEPEF. In other words, it is completely within the competence of biogeography. The traditional task of biogeography is to identify the boundaries of the species ranges and find distribution patterns of the species due to geographic, biotic, and climatic factors. The ecological aspect of the species distribution analyzing in biogeography is most often associated with the concept of ecological niche in the understanding of Grinnell [23], that is, the attitude of a species to changes in environmental parameters. Unlike Hutchinson's ecological niche [24], which is determined by the properties of a species in the hyperspace of environmental factors (i.e., the ecological preferences of the species, rather than the environment), Grinell's niche is determined by environmental parameters. Changes of these parameters lead to changes of species environmental preferences. Therefore, studying the joint change of climatic factors and the numerical characteristics of the species in space, one can identify the climatic optimum and pessimum of the species.

2. Methods

To study the Sphagnum distribution on the EEPEF, 13 species were chosen, and the literature data with annotated lists of specific bryofloras from different regions (European part of the Russian Federation, the Baltic States, Ukraine, Belarus, and Moldova) were analyzed (**Figure 1**). Some dots have been chosen outside the study area (e.g., Romania, Poland, Kazakhstan, Cauacasus, and eastern mountainside of Ural) to correct possible errors of extrapolation at the boundaries [17, 20]. Earlier,

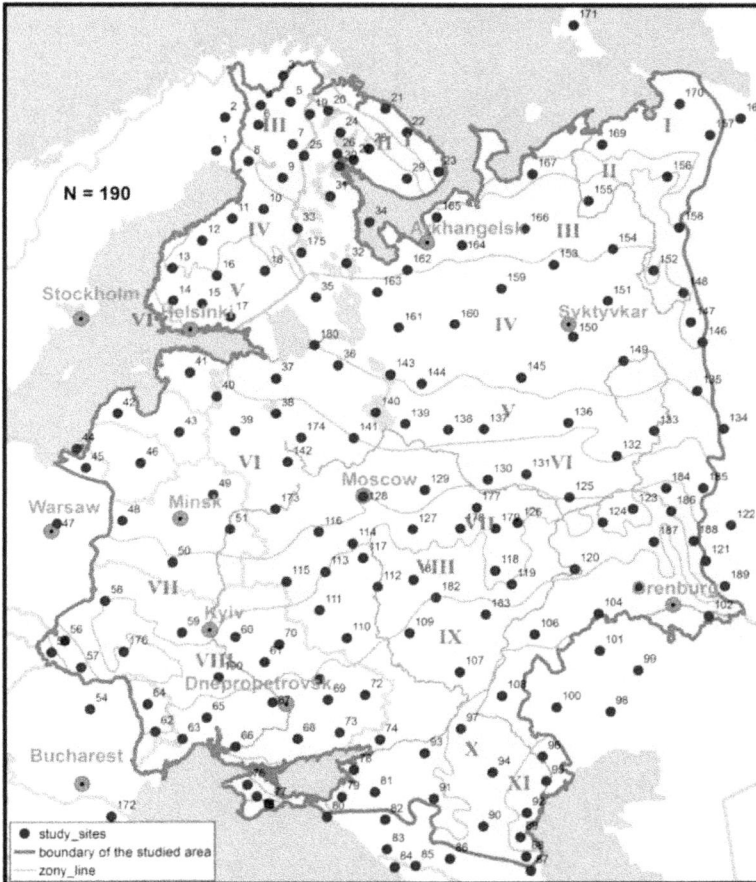

Figure 1.
Study area, showing localities involved in analysis and vegetation zones: I—Tundra; II—Forest Tundra; III—Northern Taiga; IV—Middle Taiga; V—Southern Taiga; VI—Mixed forests; VII—Broadleaved forests; VIII—Forest Steppe; IX—Steppe; X—Semidesert; XI—Desert (boundaries of vegetation zones are given by [30, 31]). Study sites: 1–18—[32]; 19—[33]; 20—[34]; 21–23—[35]; 24—[33]; 25—[36]; 26—[33]; 27–29—[35]; 30—[37]; 31–32—[38]; 33—[39]; 34—[40]; 35—[41]; 36–39—[42]; 40—[43]; 41—[44]; 42–43—[45]; 44—[46]; 45—[32]; 46—[32, 47]; 47—[48]; 48–51—[49]; 52—[50]; 53—[51]; 54—[52]; 55–57—[53]; 58–59—[54]; 60–61—[55]; 62–64—[56]; 65–69—[27]; 70–72—[57]; 73–74—[58]; 75–77—[59]; 78–80—[60]; 81—[61]; 82—[62]; 83—[63]; 84—[64]; 85—[65]; 86—[66]; 87—[67]; 88—[68]; 89–92—[61]; 93–96—[69]; 97—[70]; 98–105—[71]; 106–108—[70]; 109–115—[57]; 116—[72]; 117—[73]; 118—[74]; 119—[75]; 120—[76]; 121—[77]; 122—[78]; 123–124—[79]; 125—[80]; 126—[81]; 127—[82]; 128—[83]; 129–130—[84]; 131—[85, 86]; 132—[87]; 133—[88]; 134—[89]; 135—[90]; 136—[91]; 137–138—[92]; 139–140—[93]; 141—[94]; 142—[95]; 143—[96]; 144—[97]; 145—[91]; 146—[98]; 147—[99]; 148—[100]; 149–158—[101]; 159–162—[40]; 163—[102]; 164–166—[103, 104]; 167—[105]; 168—[106, 107]; 169–171—[107, 108]; 172—[109]; 173—[110]; 174 —[94]; 175—[111]; 176—[112]; 177—[113]; 178—[114]; 179—[115]; 180—[116]; 181–182—[117]; 183—[75]; 184–188—[118]; 189—[77]; 190—[55].

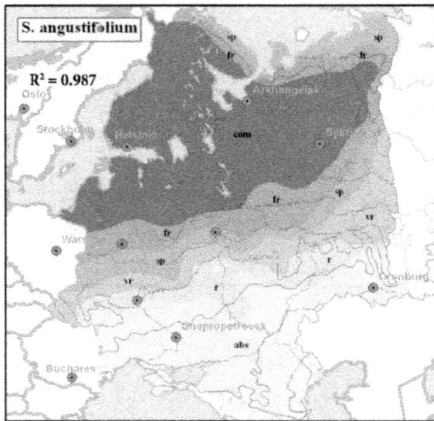

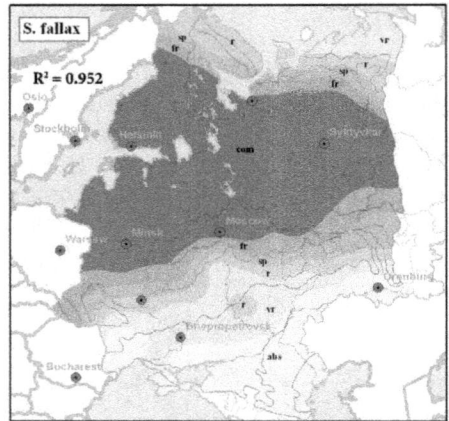

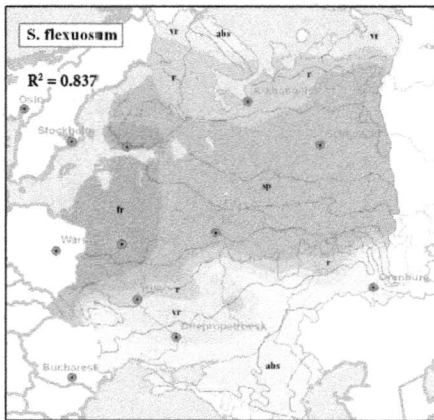

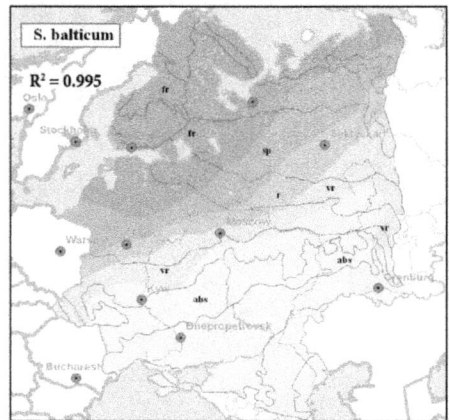

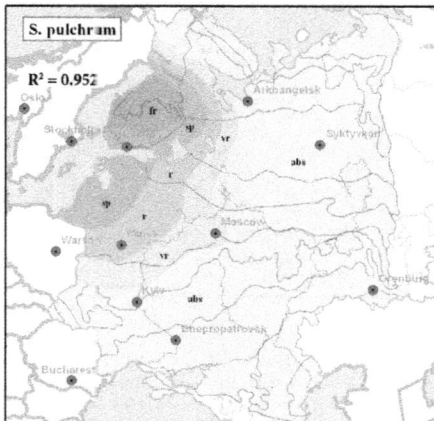

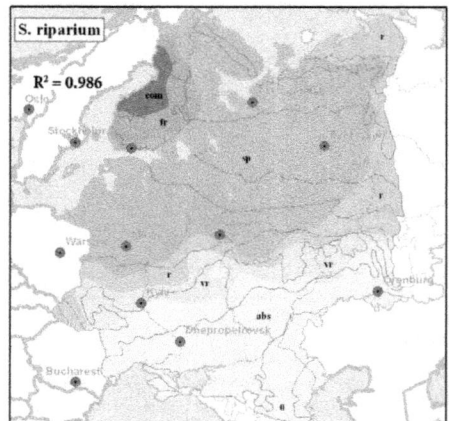

the basic principles for creating model areas by geostatistics methods using the kriging method were printed, and the methodology for compiling model maps of species ranges was adapted to the goals of biogeography [20]. After literature data compilation, the occurrence of each species was estimated in ordinal six-point scale: 0—absent (**abs**), 1—very rare (1–2 records) (**vr**), 2—rare (3–7 records) (**r**), 3—sporadically (more than 7 records, but not everywhere) (**sp**), 4—frequent (usual species, but sometimes absent in suitable phytocoenosis) (**fr**), and 5—common

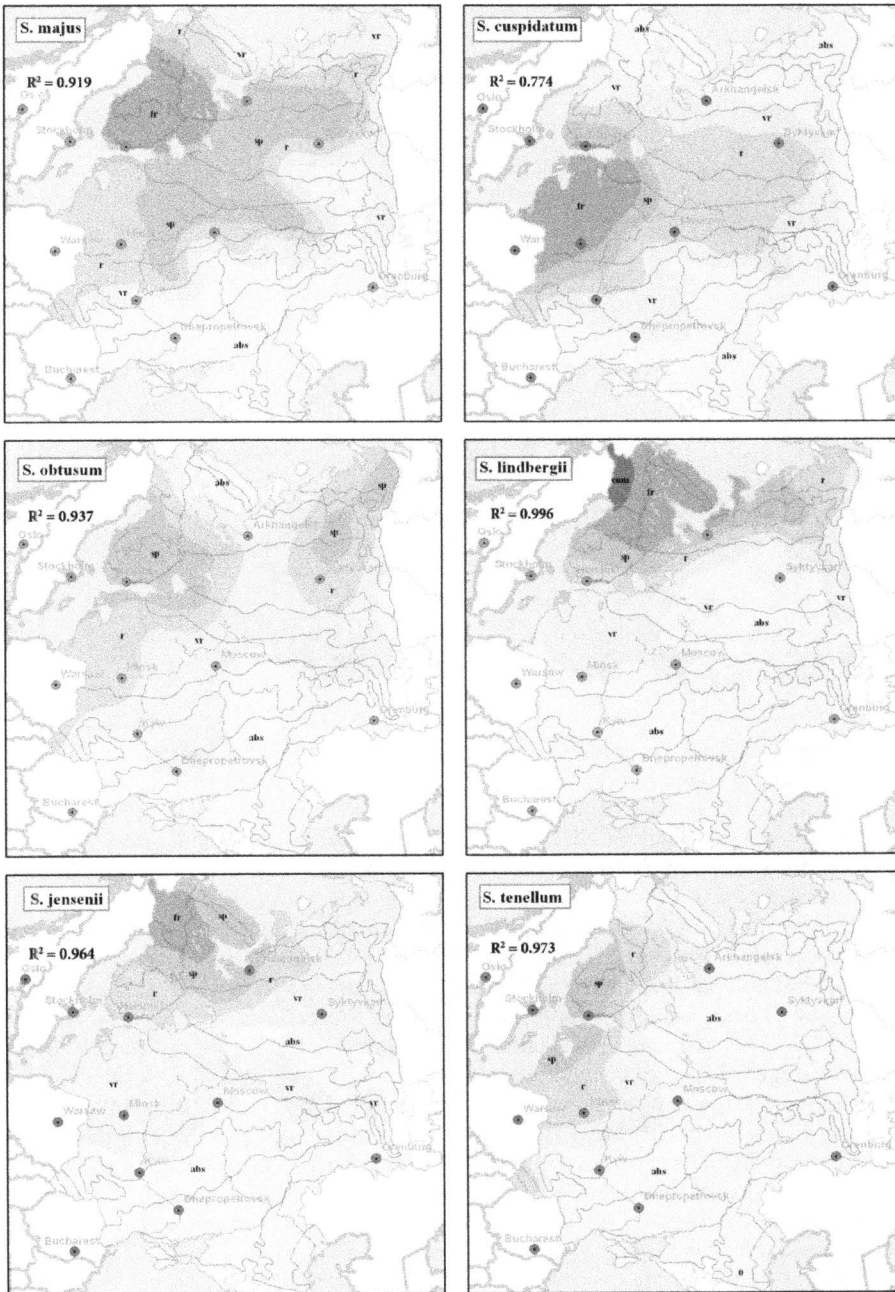

(usual and phytocenotically active species in the study area) (**com**). In the following text, these abbreviations will be used to denote areas of species occurrence. According to this scale, continuous coverages were constructed for each species using the kriging method [17] with a resolution of 10 km in 1 pixel. In total, a sample of 190 points (local floras) was used to create continuous coverages (**Figure 1**). Verification of continuous coverages was carried out by cross-validation method in the SAGA GIS software. The index of quality of cross-validation in geostatistics is

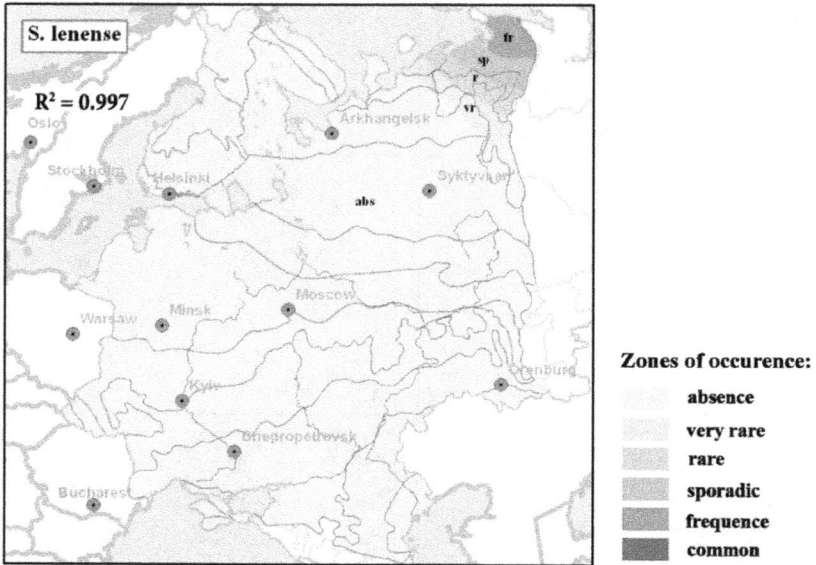

Figure 2.
*Model ranges of 13 species of the subgenus Cuspidata (the red lines indicate the boundaries of vegetation zones). Zones of occurrence: **abs**—species is absent; **vr**—very rare; **r**—rare; **sp**—sporadic; **fr**—frequent; and **com**— common. For each species on the maps is shown R².*

the coefficient of determination (R^2) [17]. The values of this indicator for continuous coverages of species under study are shown in **Figure 2**. Climatic optimum was determined for zones of frequent (**fr**) and common (**com**) occurrences.

Continuous coverages of climatic factors were used in analysis also. We used dataset that authors of WORLDCLIM program [25] propose. In total, 23 climatic variables were used. This is the following: annual mean of precipitation (**amt**), monthly temperature of April–October (**tm04–tm10**), annual precipitation (**pr_a**), monthly precipitation (**pr04–pr10**), and relative humidity (**reh04–reh10**) of April–October. We have chosen only months of growing season from dataset. Each coverage was composed in Azimuthal Equidistant Projection (Central Meridian 45°E, chief of the parallel 55°N). The coverages of climatic factors were combined with coverages of species occurrence to a single spatial database. This spatial database was converted into relative table, which contains 36 variables (23 climatic factors and 13 species occurrence) and 49,557 cases (number of pixels). This database was used for calculation of descriptive statistics and performing correlation and cluster analysis in software Statistica 10.0. Operation with creating and verification coverages was performed in SAGA software. The operations by intersection of the vector layers and calculating of areas were performed in software ArcGis 10.0. In more detail, all techniques were described in previous article [21].

3. Results

Model maps for 13 species are shown in **Figure 2**. The following species distribution patterns were found.

Sphagnum angustifolium. This species is widely distributed in the study area (**Figure 2**). The maximum score on the scale of occurrence is 5 (**com**). It grows in boggy forests and bogs. Its range is associated with the forest zone and tundra,

Zones	abs	vr	r	sp	fr	com	*Total*
Tundra			3.6	136.8	47.8	3.6	*191.8*
Forest Tundra			0.0	1.9	85.6	14.6	*102.1*
North Taiga			13.3	5.8	56.1	475.5	*550.7*
Middle Taiga			24.4	44.7	59.9	619.1	*748.0*
South Taiga		0.1	92.8	32.7	88.4	326.0	*540.0*
Mixed Forest		0.6	81.5	113.1	278.9	340.2	*814.3*
Broadleaves Forest	44.3	72.1	235.4	155.0	10.3		*517.2*
Forest Steppe	209.0	253.0	58.4	2.2			*522.7*
Steppe	659.1	49.5					*708.6*
Semidesert	204.8	0.0					*204.8*
Desert	54.9	0.0					*54.9*
Total, km²	*1172.1*	*375.4*	*509.5*	*492.1*	*626.9*	*1778.9*	*4955.0*
Total, %	*23.7*	*7.6*	*10.3*	*9.9*	*12.7*	*35.9*	*100*

Table 1.
Areas (in 1000 km²) covered by S. angustifolium by zones of its occurrence.

where such habitats are widespread. To the south of the forest zone, *S. angustifolium* decreases its abundance and completely disappears in the steppe or even in forest steppe in some places. Its occurrence increases in the northern and middle taiga. It grows in all vegetative zones from tundra to steppe (**Figure 2** and **Table 1**). The zone of its greatest occurrence (**com**) occupies 35.9% of the total area of the EEPEF. The zone of total absence is 23.7% (**Table 1**). Thus, the range of this species covers 76.3% of the total area of the EEPEF, therefore *S. angustifolium* can be considered here as a common and widespread species.

The southern boundary of the range of *S. angustifolium* (the southern boundary of **vr** zone) passes in sublatitudinal direction and is approximately parallel to the boundaries of natural zones. The border of the zone of maximum occurrence (**com**) passes diagonally to the meridians. In terms of biogeography, this is the zone of its climatic optimum. In the best way, the border of the **com** zone correlates with the boundary of the maximum occurrence of wetlands [26] and with isotherm of July +17°C and with maximal average values of air humidity in July–September.

Sphagnum fallax. This species is distributed from tundra to forest steppe zone (**Figure 2**). The maximum score on the scale of occurrence is 5 (**com**). In the south of the steppe zone, this species is absent, with the exception of its tongue with lower occurrence along Dnieper river, where it occurs on rare bogs, located on the river terraces [27]. It has maximal abundance (**com**) in the forest zone and occurs with a small abundance (**vr**) in the forest tundra and forest steppe, but here it is rare (**Table 2** and **Figure 2**). The zone of maximal occurrence of the species takes about a half area of the EEPEF (44.7%) (**Table 2**). This species is absent in 13.9% of the area only, that is, its range covers 86.1% of the EEPEF area. Thus, *S. fallax* is the most common and widespread species.

As well as *S. angustifolium*, *S. fallax* has similar climatic preferences. The boundaries of all zones of *S. fallax* are generally parallel to the boundaries of natural zones (**Figure 2**). Unlike *S. angustifolium*, *S. fallax* comes further south—its range reaches the Black Sea along the Dnieper. However, in the steppe zone, it is an extremely rare species. In the north of the EEPEF, *S. fallax* does not completely disappear, but becomes much more rare, in contrast to *S. angustifolium*, which is a fairly frequent species in the tundra (**Figure 2** and **Table 2**). The boundaries of all zones best correspond to region with the greatest average summer precipitation and air humidity, and the southern border of its range is generally well suited the isotherm of July of +21°C (southern boundary of the **vr** zone) and to +13°C in the north (southern boundary of the **r** zone) (**Figure 2**).

Sphagnum flexuosum. This species is distributed from tundra to the steppe zone (**Figure 2**). The maximum score on the scale of occurrence is 4 (**fr**). It reaches the highest occurrence (**fr**) to the west of the forest zone (**Figure 2**), but it occurs sporadically throughout almost the entire forest zone. Sporadic zone occupies most of the range of this species −41.2%—and extends from the northern taiga to the forest steppe (**Table 3**). In general, *S. flexuosum* covers 80.7% of the total area, and therefore this species, as well as two previous species, can be considered as widespread species in this area.

The boundaries of almost all zones of occurrence of this species run almost parallel to the boundaries of natural zones (**Figure 2**). The boundary of the zone **fr** passes in the submeridianal direction. This fact indicates that the optimum zone of *S. flexuosum* is limited by the factors of humidity and not by temperature. This zone is located in regions around the Baltic Sea, where relatively warm summers and the greatest amount of precipitation are observed [28]. In the south, the range of this species reaches to the northern steppes only and in the north—to the Arctic Ocean. True, in tundra, it is very rare. In the best way, the boundaries of all zones of occurrence (except for zone **fr**) correspond to the high average values of precipitation in July–September, and they have a weak correspondence with isotherms (**Table 3**).

Sphagnum balticum. The range of this species from north to south covers the area from the tundra zone to the zone of deciduous forests, and its occurrence does not exceed four on a six-point scale (**Figure 2**). In the southern Urals, it captures a small section of the forest steppe zone (**Table 4**). The maximum occurrence of *S. balticum* is observed in the tundra and in the north of the forest zone. Zone **fr** occupies about a quarter of the total area (25.6%) of the EEPEF (**Table 4**). The territory, where *S. balticum* is absent (**abs**), makes up 38.6% of the EEPEF, that is, the range of this species occupies 61.4% of total area. Therefore, *S. balticum* can also be called a relatively widespread species.

Zones	abs	vr	r	sp	fr	com	Total
Tundra		117.3	73.0	1.5	0.0	0.0	191.8
Forest Tundra		2.4	76.8	22.1	0.8	0.0	102.1
North Taiga			67.1	118.6	126.4	238.6	550.7
Middle Taiga					2.5	745.5	748.0
South Taiga				0.5	85.0	454.5	540.0
Mixed Forest				36.5	70.1	707.7	814.3
Broadleaves Forest	10.5	20.4	58.6	151.4	208.2	68.2	517.2
Forest Steppe	30.9	220.5	170.8	89.1	11.3		522.7
Steppe	390.6	289.3	28.4	0.2			708.6
Semidesert	202.5	2.2					204.8
Desert	54.9						54.9
Total, km^2	689.4	652.2	474.6	420.0	504.2	2214.5	4955.0
Total, %	13.9	13.2	9.6	8.5	10.2	44.7	100

Table 2.
Areas (in 1000 km²) covered by S. fallax by zones of its occurrence.

Zones	abs	vr	r	sp	fr	Total
Tundra	25.3	112.9	41.6	12.0		191.8
Forest Tundra	8.9	18.5	66.5	8.2		102.1
North Taiga	12.1	61.7	320.8	156.1		550.7
Middle Taiga			78.9	669.1		748.0
South Taiga			1.8	461.0	77.2	540.0
Mixed Forest		0.4	6.5	435.0	372.4	814.3
Broadleaves Forest	35.6	27.4	119.1	261.7	73.3	517.2
Forest Steppe	92.9	270.9	118.5	38.5	1.8	522.7
Steppe	524.0	174.7	9.9			708.6
Semidesert	204.8					204.8
Desert	54.9					54.9
Total, km²	*958.5*	*666.5*	*763.8*	*2041.5*	*524.7*	*4955.0*
Total, %	*19.3*	*13.5*	*15.4*	*41.2*	*10.6*	*100.0*

Table 3.
Areas (in 1000 km²) covered by S. flexuosum by zones of its occurrence.

Zones	abs	vr	r	sp	fr	Total
Tundra			5.8	33.6	152.3	191.8
Forest Tundra				23.3	78.8	102.1
North Taiga		21.3	28.9	147.6	353.0	550.7
Middle Taiga		124.2	161.2	198.5	264.0	748.0
South Taiga	57.1	110.7	74.9	77.4	219.9	540.0
Mixed Forest	92.7	187.2	139.4	195.4	199.6	814.3
Broadleaves Forest	294.2	194.3	27.6	1.1		517.2
Forest Steppe	504.4	18.3				522.7
Steppe	705.2	3.4				708.6
Semidesert	204.8					204.8
Desert	54.9					54.9
Total, km²	*1913.2*	*659.5*	*437.7*	*676.8*	*1267.8*	*4955.0*
Total, %	*38.6*	*13.3*	*8.8*	*13.7*	*25.6*	*100*

Table 4.
Areas (in 1000 km²) covered by S. balticum by zones of its occurrence.

The boundaries of the range of *S. balticum* as a whole and the boundaries of zones of occurrence within the range are oblique with respect to the borders of natural zones and show a clear tendency toward concentration around the Baltic Sea (**Figure 2**). The boundary of the zone of maximal occurrence (**fr**) lies parallel and entirely within the zone of maximum distribution of the Valdai glaciation [29]. The boundary of the zone of sporadic occurrence (**sp**) generally coincides with the zone of maximal distribution of wetlands [26]. This is not surprising if we recall that *S. balticum* is predominantly a boggy (and not forest) species, especially in the north [9–11]. Thus, it can be assumed that the distribution of *S. balticum* in the northern parts of its range, where it occurs most often, in addition to climatic factors, is influenced by the historical conditions and landscape features of the territory. The influence of climatic factors, however, also occurs, since the southern border of the **sp** zone roughly corresponds to the isotherm of July +17°C. In the best way, the boundaries of the zones of occurrence correspond to the monthly precipitation and relative humidity of air in August–September.

Sphagnum riparium. It is rather widely distributed in the EEPEF (**Figure 2**); however, in most of the area, it occurs sporadically. The **sp** zone occupies about half of the investigated area (45.8%) (**Table 5**). The maximal occurrence zone reaches in Finland and Sweden (**Figure 2**), which is connected, in my opinion, with the greater prevalence of suitable habitats in these countries, such as aapa-bogs. In general, the *S. riparium* range covers 74.7% of the EEPEF, so this species can be considered widespread in this area.

In the west, the boundary of the **sp** zone more or less coincides with the isotherm of July +17°C. In the east—in the Ural Mountains—any correspondence to climatic factors is not detected. The decrease of the occurrence of *S. riparium* in Urals seems to be due to the lack of suitable habitats.

Sphagnum majus and *S. cuspidatum.* Both species, as well as *S. riparium*, are widely distributed throughout the EEPEF, but with a small abundance. The peak of their coenotic activity is observed in western regions, where they grow jointly or separately in the flooded hollows of oligotrophic or mesotrophic bogs. Apparently, their lower occurrence in the east is related to the difference in the composition of the bog complexes of the Western European and East European bogs. Although the ranges of both species are largely similar, *S. majus* is more northern than *S. cuspidatum*. Area of *S. majus* covers 66.8% and *S. cuspidatum*—59.7% (**Tables 6** and **7**) from total area. The boundaries of the zones of occurrence of both species are weakly related to the isolines of any climatic factors, except zone **fr**. This zone (for both species) lies within the region with the highest humidity and precipitation in August–September.

Sphagnum jensenii. The boundary of the range of this species has a fancy pattern. In general, it occupies 56.1% of the total area (**Table 8**). It is most prevalent in Fennoscandia and in Russian North (**Figure 2**). Throughout its range, *S. jensenii*

Zones	abs	vr	r	sp	fr	com	Total
Tundra		2.1	128.3	61.3			191.8
Forest Tundra			13.3	88.8			102.1
North Taiga			40.4	444.1	60.0	6.2	550.7
Middle Taiga			49.4	592.6	41.5	64.5	748.0
South Taiga			110.8	323.3	90.1	15.8	540.0
Mixed Forest		6.5	128.5	672.7	6.7		814.3
Broadleaves Forest	34.6	205.9	190.9	85.7			517.2
Forest Steppe	279.5	228.3	14.9				522.7
Steppe	699.8	8.8	0.0				708.6
Semidesert	204.8	0.0					204.8
Desert	54.9						54.9
Total, km^2	1273.5	451.5	676.5	2268.6	198.3	86.6	4955.0
Total, %	25.7	9.1	13.7	45.8	4.0	1.7	100

Table 5.
Areas (in 1000 km²) covered by S. riparium by zones of its occurrence.

Zones	abs	vr	r	sp	fr	Total
Tundra		180.9	10.8	0.0	0.0	191.8
Forest Tundra		49.6	52.1	0.4	0.0	102.1
North Taiga		60.4	200.7	193.1	96.6	550.7
Middle Taiga	62.3	107.4	98.9	332.3	147.1	748.0
South Taiga	104.3	53.2	39.4	207.6	135.4	540.0
Mixed Forest	31.0	73.1	401.0	304.5	4.7	814.3
Broadleaves Forest	93.4	174.1	227.5	22.2		517.2
Forest Steppe	384.5	127.8	10.4			522.7
Steppe	708.6					708.6
Semidesert	204.8					204.8
Desert	54.9					54.9
Total, km²	*1643.7*	*826.5*	*1040.7*	*1060.1*	*383.9*	*4955.0*
Total, %	*33.2*	*16.7*	*21.0*	*21.4*	*7.7*	*100.0*

Table 6.
Areas (in 1000 km²) covered by S. majus by zones of its occurrence.

Zones	abs	vr	r	sp	fr	Total
Tundra	191.8					191.8
Forest Tundra	102.1					102.1
North Taiga	405.8	144.3	0.7			550.7
Middle Taiga	83.7	265.7	396.6	2.0		748.0
South Taiga	26.1	68.1	307.7	99.5	38.6	540.0
Mixed Forest	26.4	67.4	253.8	106.3	360.4	814.3
Broadleaves Forest	90.5	131.0	209.1	63.5	23.1	517.2
Forest Steppe	184.2	310.0	25.4	3.2		522.7
Steppe	627.0	81.6				708.6
Semidesert	204.8					204.8
Desert	54.9					54.9
Total, km²	*1997.0*	*1068.1*	*1193.3*	*274.5*	*422.2*	*4955.0*
Total, %	*40.3*	*21.6*	*24.1*	*5.5*	*8.5*	*100*

Table 7.
Areas (in 1000 km²) covered by S. cuspidatum by zones of its occurrence.

practically does not change its environmental preferences—it grows everywhere in the wet hollows of oligotrophic bogs. However, on the territory of the Russian Plain, such bogs are not rare, but wet hollows are usually occupied mainly by *S. majus*. Therefore, it cannot be said that *S. jensenii* is extremely rare due to the lack of habitats in the central and eastern parts of the range. At the same time, it cannot be said that the boundaries of the zones of occurrence are associated with isolines of climatic factors. This type of range appears to be shrinking.

Sphagnum obtusum. The maximum score on the scale of occurrence for this species is 3 (sporadically). In other words, this species does not have an optimum in the study area. At the same time, the **sp** zone "goes" to EEPEF with two tongues—from

Zones	abs	vr	r	sp	fr	Total
Tundra		139.7	13.8	38.2		191.8
Forest Tundra		55.1	9.2	37.5	0.3	102.1
North Taiga		179.4	76.0	154.1	141.2	550.7
Middle Taiga	104.7	411.9	126.8	82.0	22.7	748.0
South Taiga	181.6	220.3	134.0	4.1		540.0
Mixed Forest	111.0	696.7	6.7			814.3
Broadleaves Forest	310.0	207.2				517.2
Forest Steppe	501.8	20.9				522.7
Steppe	708.6					708.6
Semidesert	204.8					204.8
Desert	54.9					54.9
Total, km²	*2177.2*	*1931.2*	*366.5*	*315.8*	*164.3*	*4955.0*
Total, %	*43.9*	*39.0*	*7.4*	*6.4*	*3.3*	*100.0*

Table 8.
Areas (in 1000 km²) covered by S. jensenii by zones of its abundance.

Zones	abs	vr	r	sp	Total
Tundra	52.0	19.7	68.5	51.5	191.8
Forest Tundra	42.8	18.8	24.8	15.7	102.1
North Taiga	108.0	175.1	198.4	69.2	550.7
Middle Taiga	12.2	267.0	286.3	182.5	748.0
South Taiga	23.0	268.6	62.5	185.9	540.0
Mixed Forest	42.8	432.2	309.7	29.6	814.3
Broadleaves Forest	180.5	274.3	62.3		517.2
Forest Steppe	289.6	233.1			522.7
Steppe	656.3	52.3			708.6
Semidesert	204.8				204.8
Desert	54.9				54.9
Total, km²	*1666.9*	*1741.1*	*1012.6*	*534.3*	*4955.0*
Total, %	*33.6*	*35.1*	*20.4*	*10.8*	*100.0*

Table 9.
Areas (in 1000 km²) covered by S. obtusum by zones of its occurrence.

Finland and Polar Urals. This is very similar to the tongues of the last glacier [29]. This is a suggestion that this species is a glacial relic. In general, the area of this species is 66.4% of the total area of the EEPEF (**Table 9**). This species does not change its ecology when geographic areas changing—everywhere it grows on quagmire along the shores of lakes or in hollows of transitional bogs and rich fens. Therefore, in our opinion, the range of this species can also be called shrinking.

Sphagnum lindbergii. This species is quite rare on the Russian Plain. Judging by the pattern of its range—it is rather Scandinavian. The area of its range is less than half of the total area (38.7%) (**Table 10**). Therefore, this species should be considered as a species with a restricted range for the EEPEF territory. The zone of

Zones	abs	vr	r	sp	fr	com	Total
Tundra		26.5	99.7	16.6	49.1		191.8
Forest Tundra		0.4	27.8	36.1	37.9		102.1
North Taiga	0.3	63.8	120.6	118.5	208.7	38.8	550.7
Middle Taiga	371.1	190.7	48.4	59.2	63.1	15.5	748.0
South Taiga	293.7	104.7	61.6	70.4	9.6		540.0
Mixed Forest	372.5	435.1	6.7				814.3
Broadleaves Forest	509.4	7.7					517.2
Forest Steppe	522.7						522.7
Steppe	708.6						708.6
Semidesert	204.8						204.8
Desert	54.9						54.9
Total, km²	3037.9	829.0	364.8	300.7	368.3	54.3	4955.0
Total, %	61.3	16.7	7.4	6.1	7.4	1.1	100.0

Table 10.
Areas (in 1000 km²) covered by S. lindbergii by zones of its occurrence.

Zones	abs	vr	r	sp	fr	Total
Tundra	191.8					191.8
Forest Tundra	102.1					102.1
North Taiga	412.7	51.7	46.0	35.5	4.8	550.7
Middle Taiga	532.4	46.3	41.2	50.8	77.3	748.0
South Taiga	300.9	42.0	52.4	66.8	77.8	540.0
Mixed Forest	232.7	170.8	228.8	182.1		814.3
Broadleaves Forest	386.2	124.3	6.6			517.2
Forest Steppe	522.5	0.2				522.7
Steppe	708.6					708.6
Semidesert	204.8					204.8
Desert	54.9					54.9
Total, km²	3649.5	435.4	374.9	335.2	160.0	4955.0
Total, %	73.7	8.8	7.6	6.8	3.2	100.0

Table 11.
Areas (in 1000 km²) covered by S. pulchrum by zones of its abundance.

its maximal distribution occurs in subarctic regions, where an air humidity is high during the growing season (**Table 11**).

Sphagnum pulchrum and *S. tenellum*. These two species go to the EEPEF from Western Europe and Scandinavia. In the investigated area, they have a restricted range. Thus, the area occupied by *S. pulchrum* is only 26.3% and by *S. tenellum*— 24.7% of the total EEPEF area (**Tables 11** and **12**).

Sphagnum lenense. This species is widespread in Siberia and goes to the EEPEF from the Polar Urals. Also, this is a species with a restricted range. The area of its range is only 5.1% of the total area (**Table 13**). In the tundra and forest tundra, it grows on the hummocks of raised bogs. Under the conditions of the Russian plain, in forest zone, such habitats are usually occupied by *Sphagnum fuscum*. The

Zones	abs	vr	r	sp	Total
Tundra	191.8				191.8
Forest Tundra	96.5	5.5			102.1
North Taiga	341.1	131.5	78.2		550.7
Middle Taiga	617.6	26.2	50.3	54.0	748.0
South Taiga	335.1	52.8	73.4	78.7	540.0
Mixed Forest	282.1	215.8	234.0	82.4	814.3
Broadleaves Forest	383.0	112.9	21.2		517.2
Forest Steppe	515.1	7.6			522.7
Steppe	708.6				708.6
Semidesert	204.8				204.8
Desert	54.9				54.9
Total, km²	3730.5	552.2	457.1	215.1	4955.0
Total, %	75.3	11.1	9.2	4.3	100.0

Table 12.
Areas (in 1000 km²) covered by S. tenellum by zones of its occurrence.

Zones	abs	vr	r	sp	fr	Total
Tundra	51.3	13.6	29.7	50.4	46.8	191.8
Forest Tundra	56.4	15.5	20.9	9.4		102.1
North Taiga	484.6	53.2	13.0			550.7
Middle Taiga	748.0					748.0
South Taiga	540.0					540.0
Mixed Forest	814.3					814.3
Broadleaves Forest	517.2					517.2
Forest Steppe	522.7					522.7
Steppe	708.6					708.6
Semidesert	204.8					204.8
Desert	54.9					54.9
Total, km²	4702.6	82.2	63.6	59.8	46.8	4955.0
Total, %	94.9	1.7	1.3	1.2	0.9	100.0

Table 13.
Areas (in 1000 km²) covered by S. lenense by zones of its occurrence.

southern boundary of the **sp** zone approximately corresponds to the isotherm of annual mean temperature of −4°C and an annual precipitation amount of 500 mm.

4. Discussion

If we consider the model maps of species, constructed according to the value of occurrence, as their geographical range in the territory of the EEPEF, then the areas of occurrence identified on it indicate areas where mosses have optimal and pessimal conditions. Results show that almost all species have an optimal area in the regions around the Baltic Sea or in the subarctic, where the wettest conditions are observed in the EEPEF. If we express the values of the moisture factors necessary for the successful distribution of species, in absolute values, they look as follows: annual precipitation is not less than 550 mm and relative humidity is not less than 60–70%.

A total of 7 species of 13 are widespread in the study area. These are *S. angustifolium*, *S. fallax*, *S. flexuosum*, *S. balticum*, *S. riparium*, *S. majus*, and *S. cuspidatum*. All of them play an important phytocenotic role in wetlands. Restricted species have western distribution. And only *S. lenense* comes to the north of the European part of Russia from the east. Some of the restricted species, such as *S. obtusum* and *S. tenellum*, do not have an optimum in the EEPEF. This suggests that they come here only at the edge of the range, and the center of their distribution is outside the EEPEF. Abovementioned seven species are characterized by the largest phytocenotic significance in wetland communities. If we compare their ranges (**Figure 2**), it is clear that they overlap significantly, but, nevertheless, each species is characterized by its own characteristics. The *S. flexuosum* area pattern is the most different from the others. This species is practically absent in the tundra and reduces its abundance to the north and south of the forest zone. At the same time, it cannot be called the most "southern" of all seven species, since the range of *S. fallax*, for example, goes even further south than *S. flexuosum* (**Figure 2**). At the same time, *S. fallax* is able to grow in the tundra, that is, far north than *S. flexuosum*. Although *S. flexuosum* grows throughout the entire EEPEF forest zone, it is obvious that its western regions are under heavy rainfall conditions. The range of *S. angustifolium* in the southern part is similar to the pattern of the ranges of *S. fallax* and *S. flexuosum*. In the north, *S. angustifolium* comes much farther into the tundra and can be found there quite often, unlike the last two (**Figure 2**). The most northern species, perhaps, can be called *S. balticum*. On the southern limit of its range, it is limited to the southern boundary of the forest zone, and in the north, it is widely represented in taiga and in tundra. The orientation of the boundaries of its range is parallel to the boundary of the last glaciation and the zone of maximal spread of wetlands (and not the boundaries of natural zones). Such orientation of boundaries indicates that its distribution in the EEPEF is caused not only by climate parameters but also by the landscape structures that formed on the plain as they recede the glacier. This equally applies to *S. riparium*.

As correlation analysis shows (**Table 14**), the occurrence in the local floras of all species of the subgenus Cuspidata, except for *S. lenense*, *S. pulchrum*, and *S. tenellum*, has a high positive relationship with the rainfall of August (**pr08**), September (**pr09**), and October (**pr10**) (**Table 2**). According to WorldClim data [25], the maximum humidity in the EEPEF is observed in the west of forest zone and tundra zone during the summer-autumn season and sharply decreases in values starting from the south of the forest steppe zone, which is associated with an increase in monthly and average annual temperatures. Therefore, in the south, species of the subgenus Cuspidata quickly reduce their abundance, completely disappearing in the south of the steppe zone, or even further north (**Figure 2**). This is associated with high negative correlation coefficients between the values of occurrence and monthly temperatures (**tm**) of the vegetation period (**Table 14**). In the north, in the tundra, the occurrence of many species decreases, but not as sharply as at the southern limit of distribution. Apparently, despite the cold summer, they still find enough moisture here to grow successfully.

The cluster analysis conducted for 13 species of the subgenus Cuspidata by the values of 23 climatic factors shows that the studied species are divided into four clusters according to their climatic preferences (**Figure 3**). **First cluster:** *S. lenense*; **second cluster:** *S. tenellum*, *S. pulchrum*, and *S. lindbergii*; **third cluster:** *S. jensenii*, *S. obtusum*, *S. majus*, *S. cuspidatum*, and *S. balticum*; and **fourth cluster:** *S. fallax*, *S. flexuosum*, *S. angustifolium*, and *S. riparium*. It is interesting to note that within these groups, there is a similarity in environmental preferences also. So the species

Factor	ang	fal	flex	balt	cusp	jens	lenens
amt	**−0.65**	**−0.47**	−0.28	**−0.62**	0.08	**−0.66**	**−0.50**
pr04	0.05	0.30	0.41	−0.08	**0.54**	−0.08	−0.20
pr05	0.11	0.39	0.43	−0.06	**0.52**	−0.07	−0.21
pr06	0.15	0.43	**0.50**	0.00	**0.61**	0.02	−0.19
pr07	0.40	**0.65**	**0.69**	0.22	**0.71**	0.23	−0.17
pr08	**0.79**	**0.81**	**0.72**	**0.72**	**0.62**	**0.70**	−0.08
pr09	**0.80**	**0.80**	**0.74**	**0.71**	**0.50**	**0.69**	0.01
pr10	**0.77**	**0.78**	**0.67**	**0.59**	0.47	**0.58**	−0.10
pr_a	0.47	**0.69**	**0.73**	0.33	**0.72**	0.30	−0.20
reh04	**0.75**	0.48	0.39	**0.84**	0.19	**0.79**	0.25
reh05	**0.57**	0.32	0.30	**0.72**	0.17	**0.65**	0.24
reh06	**0.54**	0.43	0.45	**0.65**	0.44	**0.56**	0.15
reh07	**0.71**	**0.61**	**0.59**	**0.77**	**0.54**	**0.69**	0.05
reh08	**0.83**	**0.61**	**0.50**	**0.90**	0.30	**0.84**	0.17
reh09	**0.85**	**0.64**	**0.49**	**0.90**	0.28	**0.86**	0.16
reh10	**0.83**	**0.66**	**0.50**	**0.80**	0.22	**0.78**	0.19
tm04	**−0.75**	**−0.54**	−0.36	**−0.75**	−0.02	**−0.77**	−0.41
tm05	**−0.79**	**−0.58**	−0.42	**−0.81**	−0.09	**−0.82**	**−0.51**
tm06	**−0.81**	**−0.59**	−0.43	**−0.85**	−0.13	**−0.83**	**−0.51**
tm07	**−0.82**	**−0.59**	−0.45	**−0.89**	−0.19	**−0.86**	**−0.51**
tm08	**−0.79**	**−0.59**	−0.42	**−0.82**	−0.10	**−0.82**	**−0.50**
tm09	**−0.76**	**−0.57**	−0.38	**−0.75**	−0.05	**−0.78**	−0.40
tm10	**−0.59**	−0.44	−0.25	**−0.54**	0.11	**−0.59**	−0.20

Factor	lindb	maj	obtus	pulch	rip	tenell
amt	**−0.57**	−0.49	**−0.51**	0.02	**−0.59**	0.03
pr04	−0.21	0.12	0.19	0.27	0.02	0.15
pr05	−0.22	0.15	0.14	0.19	0.12	0.09
pr06	−0.16	0.23	0.19	0.28	0.18	0.18
pr07	0.02	0.43	0.40	0.41	0.45	0.26
pr08	**0.53**	**0.79**	**0.76**	**0.66**	**0.74**	**0.50**
pr09	**0.53**	**0.70**	**0.79**	0.43	**0.76**	0.34
pr10	0.37	**0.62**	**0.65**	0.28	**0.72**	0.16
pr_a	0.13	0.48	**0.52**	0.48	0.47	0.35
reh04	**0.75**	**0.69**	**0.69**	0.44	**0.69**	0.34
reh05	**0.63**	**0.55**	**0.53**	0.42	**0.52**	0.37
reh06	0.48	**0.56**	0.49	**0.51**	**0.51**	0.38
reh07	**0.56**	**0.70**	**0.60**	**0.58**	**0.67**	0.44
reh08	**0.75**	**0.75**	**0.72**	0.44	**0.77**	0.34
reh09	**0.76**	**0.78**	**0.75**	0.43	**0.81**	0.32
reh10	**0.66**	**0.72**	**0.73**	0.26	**0.75**	0.17
tm04	**−0.68**	**−0.61**	**−0.62**	−0.13	**−0.69**	−0.10

tm05	−0.72	−0.65	−0.67	−0.21	−0.73	−0.17
tm06	−0.74	−0.69	−0.69	−0.27	−0.75	−0.22
tm07	−0.79	−0.73	−0.70	−0.40	−0.77	−0.32
tm08	−0.72	−0.66	−0.66	−0.23	−0.74	−0.18
tm09	−0.67	−0.62	−0.61	−0.14	−0.71	−0.10
tm10	−0.50	−0.42	−0.46	0.10	−0.54	0.09

Values of r >0.5 in absolute value are highlighted in bold. All values are statistically significant at p < 0.05.
Note: Climatic factors: **amt**—annual amount of precipitation; **pr01–pr12**—monthly amount of precipitation in January–December; **pr_a**—annual precipitation average; **reh4-reh10**—relative humidity in April–October; and **tm04–tm10**—monthly temperature average in April–October.

Table 14.
The Spearmen correlation coefficient between the values of climatic factors and species abundance.

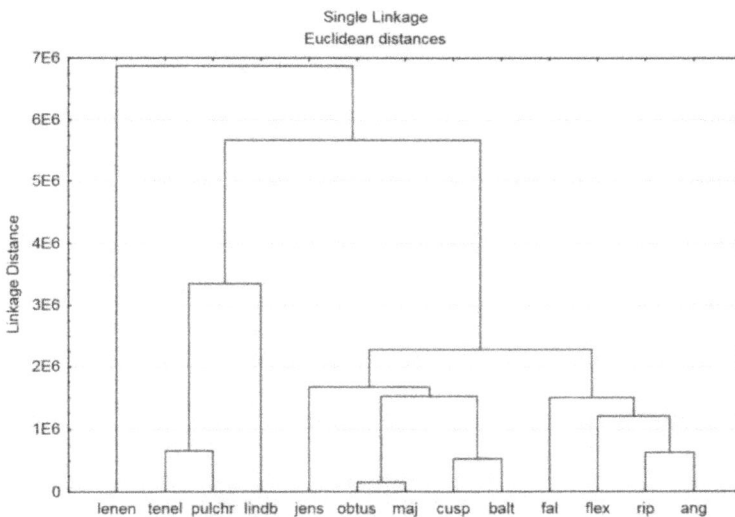

Figure 3.
Tree diagram of 13 species by 23 climatic factors.

belonging to the 4th cluster grow mainly in the carpets of mesotrophic or oligo-trophic bogs and boggy forests; the species belonging to the third cluster are most often found in heavily flooded hollows of bogs and fens; the species belonging to the second cluster grow in less flooded hollows of bogs. The S. lenense stands apart, which is found in the studied territory on hummocks in the boggy tundra. It seems to us that the similarity in the species relation to the conditions of watering of the habitat and climate is not accidental. The fact is that the amount of precipitation determines the hydrological regime in peat, and the humidity of the air affects the safety of the growing point during the dry season in the middle of summer.

5. Conclusion

Comparing the distribution ranges of 13 species of the subgenus Cuspidata in the EEPEF shows that there are as well as widespread and restricted species. The widespread species are as follows: S. angustifolium, S. fallax, S. flexuosum, S. balti-cum, S. riparium, S. majus, and S. cuspidatum. The restricted ones are S. pulchrum,

S. obtusum, *S. jensenii*, *S. tenellum*, *S. lindbergii*, and *S. lenense*. Widespread species are common in wetland communities through entire area of the EEPEF in forest zone and tundra (except *S. cuspidatum*, which is absent in tundra). Restricted species (except *S. lenense*) have western trend in its ranges. Maximum activity (optimum) of these species depends on moisture factors (humidity and precipitations), and southern boundaries are limited by temperature. The only *S. lenense* is eastern (Siberian) species. It mainly occurs in tundras and one can see a middle dependence of its distribution on the temperature factors (**Table 14**).

Author details

Sergei Yu. Popov
Lomonosov Moscow State University, Moscow, Russia

*Address all correspondence to: sergei.popov.2015@yandex.ru

IntechOpen

References

[1] Clymo RS. Experiment on breakdown of Sphagnum in two bogs. Journal of Ecology. 1965;**53**:747-757

[2] Vitt DH, Crum H, Snider JA. The vertical zonation of Sphagnum species in hummock-hollow complexes in Northern Michigan. Michigan Botanist. 1975;**14**:190-200

[3] Clymo RS, Hayward PM. The ecology of Sphagnum. In: Smith AJE, editor. Bryophyte Ecology. London: Chapman & Hall; 1982. pp. 229-289

[4] Rochefort L, Vitt DH, Bayley SE. Growth, production, and decomposition dinamics of Sphagnum under natural and experimentally acidified conditions. Ecology. 1990;**71**(5):1986-2000

[5] Vitt DH, Wai-Lin C. The relationships of vegetation to surface water chemistry and peat chemistry in fens of Alberta, Canada. Vegetatio. 1990;**89**:87-106

[6] Vitt DH. Peatlands: Ecosystems dominated by bryophytes. In: Shaw AJ, Goffinet B, editors. Bryophyte Biology. Cambridge: University Press; 2000. pp. 312-343

[7] Rydin H, Gunnarsson U, Sundberg S. The role of Sphagnum in peatland development and persistence. In: Boreal Peatland Ecosystems, Ecological Studies. Vol. 188. Berlin: Spinger-Verlag; 2006. pp. 49-65

[8] Smolyanitzkiy LY. Some regularities of the formation of Sphagnum moss cushions. Botanicheskiy Zhurnal. 1977;**52**(9):1269-1272

[9] Maksimov AI. Ecology of several peat mosses in Karelia and their role in plant communities. In: Ekologo-biologicheskie osobennosti i produktivnost' rastenii bolot. Petrozavodsk. 1982. pp. 187-195

[10] Popov SY, Fedosov VE. Coenotic distribution and ecological preferences of Sphagna in Nothern Taiga, European Russia (Pinega State Reserve, Arkhangelsk Province). Trudy KarNC. 2017;**9**:3-29. DOI: 10.17076/eco610

[11] Smagin VA, Noskova MG, Antipin VK, Boichuk MA. Diversity and phytosociological role of mosses in mires of southwestern Arkhangelsk Region and adjacent territories. Trudy KarNC. 2017;**1**:75-96. DOI: 10.17076/bg382

[12] Séneca A, Söderström L. Species richness and distribution ranges of European Sphagnum. Folia Cryptog. Estonica. 2008;**44**:125-130

[13] Geffert JL, Frahm J-P, Barthlott W, Mutke J. Global moss diversity: Spatial and taxonomic patterns of species richness. Journal of Bryology. 2013;**35**(1):1-11

[14] Hill MO, Bell N, Bruggeman-Nannenga MA, Brugués M, Cano MJ, Enroth J, et al. An annotated checklist of the mosses of Europe and Macaronesia. Journal of Bryology. 2006;**28**:198-267

[15] Isaaks EH, Srivastava RM. An Introduction to Applied Geostatistics. Vol. 561. New York: Oxford University Press; 1989

[16] Cressie NAC. The origins of kriging. Mathematical Geology. 1990;**22**:239-252

[17] Dem'yanov VV, Savel'eva EA. Geostatistics: Theory and Practice. Vol. 327. Moscow: Nauka; 2010

[18] Lur'e IK. Geoinformatical Mapping. The Methods of Geoinformatics and Digital Processing of Satellite Images. Vol. 424. Moscow: KDU; 2010

[19] Savel'ev AA, Mukharamova SS, Pilyugin AG, Chizhikova NA. Geostatistical data analysis in ecology and nature management (using the R package). Vol. 120. Kazan': Kazanskii Universitet; 2012

[20] Popov SY. Modeling the species distribution range based on the geostatistical techniques (on example of Sphagnum mosses). Trudy KarNC. 2017;**6**:70-83. DOI: 10.17076/bg558

[21] Popov SY. Distribution pattern of seven Polytrichum species in the East European Plain and Eastern Fennoscandia. Botanica Pacifica. 2018;**7**(1):25-40. DOI: 10.17581/bp.2018.07108

[22] Flatberg KI. Taxonomy of Sphagnum annulatum and related species. Annales Botanici Fennici. 1988;**25**(4):303-350

[23] Grinnel J. The niche-relationships of the California Thrasher. Foundation of Ecology. Chicago: The Univ. of Chicago Press; 1991. 118-125 p

[24] Hutchinson GE. The nicheAn abstractly inhabited hyper-volume. In: The Ecological Theater and the Evolutionar Play. New Haven: Yale Univ. Press; 1965. pp. 26-78

[25] BIOCLIM project [Internet]. 2009. Available from: http://www.andra.fr/bioclim [Accessed: 2019-01-14]

[26] Mazing V, Svirezhev YM, Loffler H, Patten BC. Wetlands in the biosphere. Wetlands and Shallow Continental Water Bodies. 1990;**1**:313-344

[27] Boiko MF. Bryobionta of the Steppe Zone of Ukraine. Vol. 264. Kherson: Ailant; 2009

[28] Alisov BP. The climat of the USSR. Moscow: Izdatel'stvo MGU; 1956. 126 p

[29] Kvasov DD. Late Quaternary History of Large Lakes and Inland Seas of Eastern Europe. Leningrad: Nauka; 1974. 278 p

[30] Ahti T, Hämet-Ahti L, Jalas J. Vegetation zones and their sections in northwestern Europe. Annales Botanici Fennici. 1968;**5**:169-211

[31] Kurnaev SF. Forest Growth Zoning of the USSR. Moscow: Nauka; 1973. 203 p

[32] Söderström L, editor. Preliminary Distribution Maps of Bryophytes in Northwestern Europe. Musci J-Z. Vol. 3. Trondheim: Mossornas Vänner; 1998. 72 p

[33] Belkina OA, Likhachev AY. Mosses of Kandalaksha State Nature Reserve (White Sea). Apaptity: Kola Science Centre; 1997. 46 p

[34] Likhachev AY, Belkina OA. Mosses of Lavna-Tundra Mountains (Murmansk Province, Russia). Arctoa. 1999;**8**:5-16. DOI: 10.15298/arctoa.08.02

[35] Shlyakov RN, Konstantinova NA. Check-list of Mosses of Murmansk Province. Vol. 227. Apaptity: Kol'skiy filial AN SSSR; 1982

[36] Drugova TP, Belkina OA, Likhachev AY. Mosses of surroundings of Alakurttii settlement and Kutsa nature reserve (Murmansk Province, North-West Russia). Arctoa. 2017;**26**(1):72-80. DOI: 10.15298/arctoa.26.07

[37] Bogdanova NE. Bryophytes of Velikiy Island (White Sea). In: Floristic researches in the Nature Reserves of the USSR. Moscow; 1981. 112 p

[38] Abramov II, Volkova LA. Handbook of Mosses of Karelia. Moscow: KMK; 1998. 390 p

[39] Boichuk MA. On the moss flora jf the Kostomuksha State Reserve and the vicinities of Kostomuksha town

(Karelia). Novosti sistematiki nizshih rastenii. 2001;**35**:217-229

[40] Churakova EY. Mosses of the Taiga zone of the Arkhangelsk Province (Northern European Russia). Arctoa. 2002;**11**:351-392. DOI: 10.15298/arctoa.11.24

[41] Boichuk MA. Mosses of Protected Areas of Karelia [thesis]. Petrozavodsk: Institute of Biology; 2002

[42] Kurbatova LV. Mosses of the Leningrad Province [thesis]. S.-Peterburg: Komarov Botanical Institute; 2002

[43] Andreeva EN, Filip'eva EO. Bryophyta of the Remda Reservation (Pskov Region). Novosti sistematiki nizshih rastenii. 2005;**38**:307-327

[44] Vellak K, Ingerpuu N, Leis M, Ehrlich L. Annotated check-list of Estonian bryophytes. Folia Cryptogamica Estonica. 2015;**52**:109-127

[45] Abolin AA. Mosses of Latvian SSR. Riga: Zinatne; 1968. 329 p

[46] Dolnik C, Napreenko MG. The bryophytes of the Southern Curonian Spit (Baltic Sea coast). Arctoa. 2007;**16**:35-46. DOI: 10.15298/arctoa.16.05

[47] Strazdiņa L, Madžule L, Brümelis G. A contribution to the bryoflora of Moricsala island Nature Reserve, Latvia. Folia Cryptogamica Estonica. 2011;**48**:107-117

[48] Stebel A. Preliminary studies on the bryoflora of Narwianski National Park (NE Poland). The Journal of Silezian Museum in Opava. 2012;**61**:265-271

[49] Rykovsky GF, Maslovsky OM. The Flora of Belarus. Bryophytes. 1. Andraeopsida–Bryopsida. Minsk: Belaruska Nauka; 2009. 437 p

[50] Dite D, Hajek M, Hajkova P. Formal definition of Slovackian mire plant associations and their application in regional research. Biologia. 2007;**62**(4):400-408

[51] Papp B, Erzberger P, Odor P, Zs H, Szoveyi P, Szurdoki E, et al. Updated checklist and red list of Hungarian Bryophytes. Studia Botanica Hungarica. 2010;**41**:31-59

[52] Erzberger P, Hohn M, Pocks T. Contribution to the bryoflora of Câlimani Mountains in the Eastern Carpatians, Romania. Acta Biologica Plantarum Agriensis. 2012;**2**:73-95

[53] Zerov DK, LYa P. The Bryophytes of Ukrainian Carpathians. Kiev: Naukova Dumka; 1975. 230 p

[54] Lazarenko AS. Handbook of Mosses of Ukraina. Kiev: AN Ukrainy; 1955. 467 p

[55] Gapon SV. Check-list of bryoflora of Ukrainian Leftbank Forest-Steppe. Poltava: Poltavsky Derzhavny Institut; 1997. 37 p

[56] Simonov GP. Handbook of mosses of Moldavian SSR. Kishinev: Shtiinza; 1978. 168 p

[57] Popova NN. Bryoflora of the Central Russian Uppland. I. Arctoa. 2002;**11**:101-168. DOI: 10.15298/arctoa.11.12

[58] Sereda VA, Ignatov MS. Bryoflora of Nothern Azov area (Rostov-On-Don Province, European Russia). Arctoa. 2008;**17**:185-190. DOI: 10.15298/arctoa.17.15

[59] LYa P. The bryoflora of Crimea. Kiev: Fitosociocentr; 2005. 170 p

[60] Ignatova EA, Ignatov MS, Seregin AP, Akatova TV, Konstantinova NA. Bryophyte flora of the projected Utrish Reserve (North–West Caucasus,

Russia). Arctoa. 2005;**14**:39-48. DOI: 10.15298/arctoa.14.04

[61] GYa D. Mosses of the Southern Kalmykia (European Part of Russia). Novosti sistematiki nizshih rastenii. 2011;**45**:292-300

[62] Akatova TV, Ignatova EA. On the moss flora of Lagonaki Highland (Adygea Rebublic, Western Caucasus). Arctoa. 2015;**24**:148-155. DOI: 10.15298/arctoa.24.15

[63] Akatova TV. Moss flora of the Caucasian Nature Reserve (Western Caucasus, Russia). Arctoa. 2002;**11**: 179-204. DOI: 10.15298/arctoa.11.15

[64] GYa D. The mosses (Bryophyta) of Abkhazia. Novosti sistematiki nizshih rastenii. 2015;**49**:295-313

[65] Ignatova EA, Ignatov MS, Konstantinova NA, Zolotov VI, Onipchenko VG. Moss flora of Teberda Reserve. Moscow: Nauka; 2008. 86 p

[66] Kharzinov Z, Portenier N, Ignatova E, Shagapsoev S, Ignatov M. Rare species and preliminary list of mosses of the Kabardino-Balkaria (Caucasus). Arctoa. 2004;**13**:33-40. DOI: 10.15298/arctoa.13.05

[67] Abakarova AS, Fedosov VE, Doroshina GY. Mosses of Tsudakhar (Dagestan, Caucasus). Arctoa. 2015;**24**:536-540. DOI: 10.15298/arctoa.24.45

[68] Ignatov MS, Fedosov VE, Ignatova EA. Moss flora of Gunib Area in Dagestan, Eastern Caucasus. Arctoa. 2010;**19**:87-96. DOI: 10.15298/arctoa.19.07

[69] Suragina SA, Ignatova EA, Ignatov MS. Contribution to the moss flora of Astrakhan Province (South European Russia). Arctoa. 2002;**10**:169-174. DOI: 10.15298/arctoa.11.13

[70] Suragina SA. Mosses of the Volgograd Province (South-Eastern European Russia). Arctoa. 2001;**10**: 45-70. DOI: 10.15298/arctoa.10.06

[71] Spirina UN, Zolotov VI. Mosses of the Orenburg State Nature Reserve (South-Eastern European Russia). Arctoa. 2004;**13**:51-56. DOI: 10.15298/arctoa.13.07

[72] Teleganova VV. Mosses of Kaluga (Middle European Russia) and their reproductive features. Arctoa. 2008;**17**:169-184. DOI: 10.15298/arctoa.17.14

[73] Popova NN, Teleganova VV, Boychuk MA. Bryoflora of the memorial and nature museum-reserve "Kulikovo pole" (Tula Province, Middle European Russia). Arctoa. 2015;**24**(2):567-573. DOI: 10.15298/arctoa.24.49

[74] Serebryakova NN. Ecological and biological features of mosses and their use in environmental monitoring [thesis]. Saratov: State Research Institute of Industrial Ecology; 2009

[75] Doroshina-Ukrainskaya GYa. Bryophytes. In: Works of the State Nature Reserve «Privolzhskaya forest-steppe». Penza; 1999. p. 43-46

[76] Popov SY. Flora of peat mosses of Zhiguli State Reserve. In: Biodiversity of Protected Areas: Estimate, Conservation, Monitoring. Moscow-Samara: Zhiguli State Reserve; 2000. pp. 194-196

[77] Zolotov VI, Baisheva EZ. Moss flora of «Shulgan-Tash» Nature Reserve (Republic Bashkortostan, Russia). Arctoa. 2003;**12**:121-132. DOI: 10.15298/arctoa.12.13

[78] Baisheva EZ, Ignatova EA, Kalinauskaite N, Potemkin AD. On the Bryophyte flora of Iremel National Park (Southern Urals). Arctoa.

2015;**24**(1):194-203. DOI: 10.15298/
arctoa.24.19

[79] Ariskina NP. Mosses of Tatar ASSR: Guide. Kazan: Kazansky Universitet; 1978. 122 p

[80] Ignatov MS, Ignatova EA, Konstantinova NA. Bryophyte flora of the Volzhsko-Kamskiy Nature Reserve (Tatarstan, European Russia). Arctoa. 2005;**14**:49-66. DOI: 10.15298/ arctoa.14.05

[81] Popov SY, Moshkovsky SA, Belovezhets KI, Chupalenkova TS, Mel'nichenko NL, Ignatov MS. Contribution for the Bryophyte Expedition to Prisursky Nature State Reserve. Ecologicheskiy vestnik Chuvashskoy respubliki. 2001;**25**:29-34

[82] Volosnova LF, Ignatova EA, Ignatov MS. Bryophyte flora of Oksky Nature Reserve (European Russia, Ryazan Province). Arctoa. 2000;**9**:3-11. DOI: 10.15298/arctoa.09.02

[83] Ignatov MS, Ignatova EA, Fedosov VE, Konstantinova NA. Bryophytes of Moscow Province: A Gude. Moscow: KMK Scientific Press Ltd; 2011. 320 pp

[84] Popov SY, Fedosov VE, Moshkovsky SA, Ignatov MS. Moss flora of Kerzhensky State Reserve (Nizhniy Novgorod Province, European Russia). Arctoa. 2004;**13**:57-66. DOI: 10.15298/ arctoa.13.08

[85] Czernjadieva IV. Mosses of Bol'shaya Kokshaga State Reserve. Novosti Sistematiki Nizshyh Rastenii. 2001;**35**:266-278

[86] Czernjadieva IV, Konstantinova NA, Bogdanov GA, Popov SY. The Anthocrotae and Bryophytic in the Bolshaya Kokshaga Reserve. Nauchnye trudy zapovednika Bolshaya Kokshaga. 2013; **6**:91-119

[87] Rubtsova AV. Bryoflora of Udmurt Republic. In: Actual Problems of Bryology. S.-Peterburg: NC RAN; 2005. pp. 171-177

[88] Bezgodov AG. On the bryoflora of the Kungur city environs (Perm Province). Arctoa. 2002;**11**:53-62. DOI: 10.15298/arctoa.11.07

[89] Djachenko AP, Ignatova EA, Marina LV. Mosses of the Visimskij State Reserve (Middle Ural Mountains). Arctoa. 1996;**6**:1-6. DOI: 10.15298/arctoa.06.01

[90] Ignatova EA, Ignatov MS, Bezgodov AG. Moss flora of the Basegi State Reserve (Perm Province, Middle Ural Mountains). Arctoa. 1995;**4**:23-34. DOI: 10.15298/ arctoa.04.04

[91] Zheleznova GV. On the moss flora of Kirov Province. Arctoa. 2014;**23**: 212-218. DOI: 10.15298/arctoa.23.18

[92] Fedosov VE, Popov SY. Bryophyte flora of Kostromskaya Taiga Station (European Russia, Kostroma Province). Arctoa. 2004;**13**:183-195. DOI: 10.15298/ arctoa.13.14

[93] Volkova LA, Zhukova AL, Potemkin AD, Nemceva ND. Bryophytes of Darvin State Reserve. In: Flora and vegetation of the Tver Province, Tver'. Tverskoi Gosudarstvennyi Universitet; 1994. pp. 13-24

[94] Notov AA, Spirina UN, Ignatova EA, Ignatov MS. Mosses of the Tver' Province (Middle Part of European Russia). Arctoa. 2002;**11**:297-332. DOI: 10.15298/arctoa.11.21

[95] Ignatov MS, Ignatova EA, Kuraeva EN, Minaeva TY, Potemkin AD. Bryophyte flora of Zentral'no-Lesnoj Biosphere Nature Reserve (European Russia, Tver Province). Arctoa. 1998;**7**:45-58. DOI: 10.15298/ arctoa.07.07

[96] Karmazina EV. Ecologo-coenotic characteristics of bryophytes of the national Park «Russian North» [thesis]. Moscow: Lomonosov Moscow State University; 2013

[97] Filippov DA, Boichuk MA. The mosses of Shichengsky Reserve (Vologda Province). Vestnik severnogo (Arkticheskogo) federal'nogo universiteta. Vol. 2. Seria: Estestvennye nauki; 2015. pp. 80-89

[98] Djachenko AP, Djachenko EA. Mosses of the reserve «Denezhkin Kamen'». In: Bryology: Tradition and Modernity. S.-Peterburg: ATTASHE; 2010. pp. 59-64

[99] Ignatova EA, Ignatov MS, Bezgodov AG. Mosses of the Vishera State Reserve (Perm Province, Northern Ural Mountains). Arctoa. 1996;**6**:7-19. DOI: 10.15298/arctoa.06.02

[100] Zheleznova GV, Shubina TP. Bryoflora of the Pechora-Ilych Biosphere Reserve. In: Flora and vegetation of the Pechora-Ilych Biosphere Reserve. Ekaterinburg: UrO RAN; 1997. pp. 175-210

[101] Zheleznova GV. The moss flora of the European North-East. S.-Peterburg: Nauka; 1994. 149 p

[102] Boichuk MA, Antipin VK, Baklin VA, Lapshin PN. Contribution to the bryoflora of Vodlozero National Park. Novosti sistematiki nizshih rastenii. 2002;**36**:213-224

[103] Ignatov MS, Ignatova EA, Popov SY, Churakova EY, Braslavskaya TY, Kucherov IB. Mosses. In: Ecosystem Components and Biodiversity of Karst Areas of Russian European North-East. Arkhangelsk: GPZ «Pinezhsky»; 2008. pp. 177-197

[104] Popov SY, Buryanina NN. Ecological features of Sphagnum mosses in the Northern taiga. In: Long-term dynamics of ecosystem components of the natural complex of the Pinezhsky nature reserve and adjacent areas. Arkhangelsk: GPZ «Pinezhsky»; 2012. pp. 51-63

[105] Zheleznova GV, Shubina TP. Mosses of the Belaya River Basin (Nothern Timan, Nenets Autonomous District). Arctoa. 2015;**24**:204-209. DOI: 10.15298/arctoa.24.20

[106] Czernyadjeva IV. The moss flora of the region of Sob Station (Polar Ural). Arctoa. 1994;**3**:133-138. DOI: 10.15298/arctoa.03.08

[107] Czernyadjeva IV. Moss flora of Yamal Peninsula (West Siberian Arctic). Arctoa. 2001;**10**:121-150. DOI: 10.15298/arctoa.10.13

[108] Afonina OM, Czernyadjeva IV. Mosses of the Russian Arctic: Check–list and bibliography. Arctoa. 1995;**5**: 99-142. DOI: 10.15298/arctoa.05.07

[109] Natcheva R, Ganeva A. Check-list of the bryophytes of Bulgaria. II. Musci. Cryptogamica Bryologica. 2005;**26**:149-172

[110] Ignatov MS, Ignatova EA. Moss flora of Middle European Russia. Moscow: KMK Scientific Press Ltd; 2003. 608 p

[111] Maksimov AI, Kuznetsov OL, Maksimova TA. Moss flora of the planned national park Tulos (Republic of Karelia). Novosti sistematiki nizshih rastenii. 2009;**36**:362-376

[112] Bolyukh VA. A comparison of moss flora of Central Podolia (Ukraine) and adjacent regions. Arctoa. 1995;**4**:45-54. DOI: 10.15298/arctoa.04.06

[113] Shestakova AA. Ecologo-coenotical and floristical features of organization of bryobiota on the Nizhniy Novgorod Province [thesis]. Nizhniy Novgorod: Lobachevsky State University; 2005

[114] Czernyadjeva IV, Mežaka A, Grishutkin OG, Potemkin AD. Bryophyta of Mordosky Reseve. In: Flora and Fauna of State Reserves. Moscow: Nauka; 2017. p. 29

[115] Silaeva TB, Chugunov GG, Kirjukhin IV, Ageeva AM, Vargot EV, Grishutkina GA, et al. Flora of National Park Smolny. Mosses and vasculars. In: Flora and fauna of National Parks. Moscow: Nauka; 2011. p. 128

[116] Volkova LA, Kuzmina EO, Boch MS, Luknickaya AF, Chaplygina OY, Belyakova RN, et al. Mosses, algens and lichens of Nizhnesvirsky State Reserve. In: Flora and Fauna of State Reserves. Moscow: Nauka; 1996. p. 34

[117] Jandovka LF, Mamonova NS. Most frequent mosses in the Tambov region. TSU Bulletin. 2009;**14**(1):166-167

[118] Baisheva EZ. Sphagna in the Republik of Bashkortostan (The Southern Ural). In: Materials of the VI International Symposium "Biology of Sphagnum Mosses"; Tomsk. 2016. pp. 11-13

Chapter 4

Bryophytes: A Potential Source of Antioxidants

Dheeraj Gahtori and Preeti Chaturvedi

Abstract

A variety of degenerative diseases are caused by free radicals. Oxidative stress, the major cause of the diseases, is due to the imbalance between the free radicals and the antioxidants. To overcome this imbalance, the body needs antioxidants whether endogenously present or supplied from exogenous sources. Hence, the search of effective natural antioxidants is greatly needed to fight the onset of degenerative diseases and aging. Indeed, vascular plants are well-known sources of good and efficient natural antioxidants. Non-tracheophytes are however relatively unexplored. Interestingly, these atracheophytes are endowed with the remarkable property of desiccation tolerance which makes them unique in the plant kingdom. The property is attributed to its specialized structure and rich reservoir of phytochemicals. Therefore, there is a need to bioprospect this rich resource for antioxidants.

Keywords: bryophytes, antimicrobial, antioxidant, phytochemicals

1. Introduction

The use of medicinal plants for treating human ailments is as old as the mankind. Man's keen observations of the mother nature led to disclosure of various curative properties of plants. These properties were acquired by the plants during evolution as adaptive strategies for protecting against various abiotic and biotic challenges faced by the plants. Changing climatic conditions of the earth also played an important role in designing plant's adaptive abilities. Human has utilized these abilities of the plants for ensuring his own survival. Needless to say, before the introduction of modern medicine, disease treatment was mainly managed by herbal remedies. Plants were found to be a rich source of therapeutic agents and hence contributed to the drug industry for a long time. Today also, many important medicinal compounds are derived from plant sources. There is huge potential to further harness this resource by exploring the vast diversity present in the plant world [1].

With increase in awareness and people becoming more health conscious, their attitude toward medicine and diet has undergone a dramatic transformation. Now, there is increased focus on plant-based diet and healthcare supplements. The natural supplements are relatively healthier and free from side effects of harmful chemicals. In human body, different natural mechanisms are responsible for production of free radicals and other reactive oxygen species (ROS). These species perform dual functions, viz., lethal as well as favorable, depending upon their concentrations. The delicate equilibrium between these two contrary effects needs to be maintained for a healthy life. At low or optimum levels, reactive oxygen species exert positive effects on cellular redox signaling and immune function, but at higher concentration, they produce oxidative stress, which may be responsible for onset of many degenerative diseases, apoptosis, aging, and food rancidity [2]. Therefore, wholesome antioxidant diet or natural antioxidant supplements should be used for a healthy life. Further, the novel and nonconventional sources of these antioxidants need to be documented regularly for relaxing the dependence on traditional sources on the one hand and for utilization of potential sources in the future as well.

Earlier, the food habits of man were ensuring sufficient intake of antioxidants in the form of fresh fruits, vegetables, and spices. In the fast-food age, change in food habits led to insufficient supply of these antioxidants. Now again, a need is being felt to use more and more of antioxidants in our day-to-day diet. As of today, both synthetic and natural antioxidants are very commonly used in food industry for increasing shelf life and improving quality of food. Another major industry using these antioxidants is medicine where they are mainly used for developing dietary supplements to promote health effect. Besides, cosmetic industry and herbal thera-peutics are also using different types of natural as well as synthetic antioxidants. Needless to say, in today's scenario, the use of synthetic antioxidants is diminishing due to increasing public awareness related to their long-term carcinogenic effect which has brought about strict legislation on their use as food additives. Nowadays, natural antioxidants are increasingly being preferred over their synthetic coun-terparts. Presently, the importance of the plant-based antioxidant constituents in providing protection against deadly diseases like cancer and heart problems as well as promoting overall health is increasingly being realized all over the world [3].

Phytochemicals derived from plants are major source of antioxidants. These phytochemicals are redox-active molecules and are dynamic to maintain redox balance in the body. Undoubtedly, plant-derived natural antioxidants are supposed to have more progressive effect on the body than synthetic ones. This is because plant con-stituents are a part of physiological functions of living flora and thus well suited to the human body. In recent years, the rising importance of biologically active components of plant origin has gained increased significance as highly promising prophylactic and restorative measures to combat diseases caused by oxidative stress. Higher plants, in particular, angiosperms, are used and explored as antioxidant sources. Cryptogams, especially bryophytes, hold rich reservoir of unique phytochemicals imparting them a strong defense mechanism to survive under highly diverse habitats despite having a non-lignified structure. There is huge potential to utilize this untrapped resource in modern healthcare as eco-friendly antibiotics and antioxidants [4].

Bryophytes, including liverworts, hornworts, and mosses, are phylogenetically placed between algae and vascular plants and form a unique division in the plant kingdom. They are small, mostly terrestrial, photosynthetic, spore-bearing plants that generally require a humid environment but can be found all over the world. These are represented by ca 7266–9000 liverworts, ca 221–225 hornworts, and 12,700–13,373 mosses [5, 6]. This large diversity of bryophytes also act as a "remark-able reservoir" of natural products or secondary compounds such as terpenoids, flavonoids, alkaloids, glycosides, saponins, anthraquinons, sterols, and other aromatic compounds. Many of them show interesting biological activity and become a potential source of different medicines. They also possess anticancer and antimi-crobial activity due to their unique chemical constituents [7].

2. What are antioxidants?

The chemical reaction that can produce free radicals and leads to chain reac-tions that may damage the cells of organisms is known as the oxidation, and the compounds that inhibit or retard the oxidation of compounds are known as antioxidants. Antioxidants are broadly classified into three groups [8].

1. The first group of antioxidants is the enzymes which include catalase, superoxide dismutase, peroxidase, and glutathione reductase along with the minerals like Se, Cu, Zn, Fe, Mn, etc. that act as cofactors of these enzymes.

2. The second group of antioxidants includes glutathione, vitamin E (tocopherols), vitamin C, lipoic acid, albumin, carotenoids (vitamin A), phenolics, and flavonoids.

3. The third group of antioxidants includes a complex group of enzymes like DNA repair enzymes, transferases, lipases, proteases, methionine sulfoxide reductase, etc. which are used for repair of damaged DNA, damaged proteins, oxidized lipids, and peroxides [9].

The chemical compounds and reactions which are capable in generating potential toxic oxygen species/free radicals are referred to as "prooxidants." They attack macromolecules including proteins, DNA, and lipids and cause cellular or tissue damage. In a normal cell, due to the result of imbalance between reactive oxygen species (ROS) and antioxidant defenses, the oxidative stress is generated. It can result in serious cell damage if the stress is massive or prolonged. This leads to improper functioning which causes different pathogenic conditions like aging, carcinogenesis, cardiovascular dysfunction, neurodegenerative diseases, etc.

The reactive oxygen species (ROS) is generated during the different essential processes like photosynthesis, respiration, and stress responses. These ROS can lead to the disruption of the normal physiological and cellular functions and also the biomolecules of plasma membranes and cell walls [10, 11]. Interestingly, there are both ROS producing as well as ROS quencher systems operational in various organelles of cell. Low levels of ROS are beneficial sometimes acting as signaling molecules for stress tolerance by causing upregulation of the genes involved in the pathway of synthesis of stress enzymes/metabolites. High concentration of ROS is, however, deleterious and needs to be scavenged by either the intake of antioxidants or body's own endogenous antioxidants.

3. Antioxidant property in bryophytes

Bryophytes constitute a group of small plants which form essential components of terrestrial ecosystems. These are moisture-loving plants found mostly at the sites where water is readily available [12]. Although nowadays these plants are increasingly being focused for therapeutic research, the backbone of therapeutics, i.e., the chemistry of the group, is too limited covering less than 10% of the bryophytes [13]. Bryophytes possess good biological activities. The diverse activities of the bryophytes ranged from antimicrobial, cytotoxic, antitumor, cardiotonic, allergy causing, irritancy and tumor effecting, insect anti-feedant, molluscicidal, piscicidal, plant growth regulatory to superoxide anion radical release inhibition and 5-lipoxygenase, calmodulin, hyaluronidase, and cyclooxygenase inhibitions [14].

Among all the bryophytes, liverworts, being remarkable reservoir of natural products, are therapeutically used worldwide, especially in Indian and Chinese systems of medicine for the treatment of hepatitis and skin disorders [15–17]. Mosses, though more diverse than liverworts, are relatively lesser explored for medicinal utility. The secondary metabolites identified from mosses belong to terpenoids, flavonoids, and bibenzyls. They are also rich in other compounds such as fatty acids, acetophenols, etc. Their antimicrobial activity is related to the specific chemical composition, structural configuration of compounds, functional groups, as well as potential synergistic or antagonistic interactions between compounds [14].

Bryophytes produce a number of secondary metabolites that strengthen these delicate plants with strong antioxidative machinery to cope up with biotic and abiotic stresses [18, 19]. To compensate for the absence of any special morphological and anatomical defense mechanism, these plants have developed active molecular and chemical defenses for their protection. The antioxidant defenses provide protection to the cell membranes and cell organelles against oxidative damage. Under unfavorable conditions, reactive oxygen species react with important cell constituents, viz., proteins and lipids, causing disruption of cell structure ultimately leading to cell damage. Antioxidant enzymes protect cells against the oxidative stress induced by both internal and external unfavorable conditions. High level of these antioxidants present in liverworts and mosses can serve as a future source for medicinally and cosmetically significant compounds [20].

Several bryophytes have been reported to show significant antioxidant activity. Some of these bryophytes possessed very efficient antioxidant enzyme systems, while others showed the presence of diverse kinds of phenolics and flavonoid compounds responsible for free radical scavenging. In one such study on the liverwort *Marchantia polymorpha*, antioxidant enzyme peroxidase was characterized which was found to be different from any known peroxidase of vascular plants [21]. Similarly, a search for antioxidant enzymes in a moss, *Brachythecium velutinum*, and a liverwort, *M. polymorpha*, showed the role of an enzyme, ascorbate peroxidase, in the removal of hydrogen peroxide [22]. In another study, the extract of *Plagiochasma appendiculatum* showed significant antioxidant activity by inhibiting lipid peroxidation and increasing superoxide dismutase and catalase activity [23]. Reverse-phase high-pressure liquid chromatography reported the presence of various phenolic compounds such as caffeic, gallic, vanillic, chlorogenic, p-coumaric, 3-4 hydrozybenzoic, and salicylic acid in the moss *Sphagnum magellanicum* [24]. Other studies determined the presence of phenols, flavonoids, saponins, tannins, and glycosides in *M. polymorpha*. These studies also indicated anticancerous role of flavonoids extracted from cell suspension cultures of *M. linearis* against colon cancer cell lines [25, 26]. The biological characteristics of the terpenoids and aromatic compounds isolated from bryophytes also showed antibacterial and antifungal activities [27, 28]. Like other plants, antioxidant activity of bryophytes is influenced by several factors, viz., altitude, tissue type, and seasons [29]. The biochemical compounds responsible for antioxidant activity are also subject to quantitative and qualitative change in response to changes in these factors.

Bryophytes are traditionally used in the Chinese, Indian, and American societies for various medicinal purposes. However, the ethnomedicinal use of bryophytes needs to be scientifically investigated and validated for active principles in order to bridge the gap between traditional knowledge and pharmacology. For this, the active principle responsible for the specific activity may be identified and purified. The study on the antioxidant activities of the extracts of *Oxytegus tenuirostris*, *Eurhynchium striatum*, and *Rhynchostegium murale* showed that the climate is the most important ecological factor that determines the antioxidant property of the moss. Depending on these factors, antioxidant amounts in the species vary both within themselves and between species [30]. The study on the total free radical scavenging activity of *Eurhynchium striatulum* and *Homalothecium sericeum* showed that these have very strong free radical scavenging activity [31].

The alpine moss, *Sanionia uncinata*, produces some secondary metabolites that help the plant against the environmental stresses such as UV, drought, and high temperatures. *S. uncinata* shows good antioxidant activity, free radical scavenging activity, reducing power, superoxide radical scavenging activity, and ABTS [2,2′-azino-bis(3-ethylbenzthiazoline-6-sulfonic acid)] cation scavenging activity [32]. A study on the extracts of *Polytrichastrum alpinum* revealed that isolated compounds have two to sevenfold increased antioxidant activity than their extracts [33]. The reducing power of plant extracts was reported to be directly correlated with their antioxidant activity [34] and is based on the presence of reductones, which exert antioxidant activity by breaking the free radical chain and donating a hydrogen atom [35].

The remarkable nature of polyphenolics in terms of antioxidant potential has been identified to cure many lifestyle diseases [36]. Polyphenolic molecules contain one or many aromatic rings with hydroxyl groups. Generally, the antioxidant capacity of the phenolics is directly related with the number of free hydroxyls and conjugation of side chains with the aromatic rings [37]. The phytochemical studies on *Thuidium tamariscellum* showed the presence of significant level of terpenoids in the moss. High antioxidant property shown by the plant is reported to be mainly due to the presence of considerable amount of terpenoids [38].

Studies also revealed that the total flavonoid contents of liverworts were generally higher than those of mosses. Acrocarpous mosses had generally higher values of these compounds than that of pleurocarpous mosses. The total flavonoid contents of bryophytes growing at lower light levels were higher than those growing in full-sun. Likewise, total flavonoid contents of epiphytic bryophytes were highest, while those of aquatic bryophytes were the lowest. Species growing at low-latitudes had higher flavonoid content than those at high latitudes [39]. Studies also revealed that the antioxidant values of liverworts were closer to those of vascular plants. Guaiacol peroxidase and catalase activity of *P. appendiculatum* was found higher than *Pellia endivaefolia*, while superoxide dismutase, ascorbic acid, proline, glutathione, and total phenols were found higher in *P. endivaefolia* than *P. appendiculatum* [40].

Antioxidant and free radical scavenging activities are in the focus of attention of both medical practitioners and dieticians. Free radicals are supposed to play a key role in the pathogenesis of many diseases [41]. Oxidation processes may also decrease the stability of drugs and foods. Reactive oxygen species (ROS) and reactive nitrogen species (RNS) have been recognized as fundamental components of stress signal cascades [42] under both abiotic and biotic stresses [43, 44]. Bryophytes occupy a special position among plants because the haploid gametophyte dominates their life cycle. Some species have been studied for their tolerances to drought and water stress (flooding) [45, 46] or high nitrogen concentrations [47]. Mosses are common in the vegetation of all continents, but they are still highly marginalized in traditional medicines. The plants which can respond and adapt to drought stress are certainly better equipped with complex and highly efficient antioxidative defense systems comprising of protective nonenzymatic as well as enzymatic mechanisms that efficiently scavenge ROS and prevent damaging effects of free radicals [48].

Some crude extracts of mosses contain hypnogenols, biflavonoids, dihydroflavonols, polycyclic aromatic hydrocarbons, and hydroxy flavonoids [49–51]. Flavonoids are synthesized by plants in response to the microbial infection. This action is probably due to their ability to complex with extracellular and soluble proteins and to complex with bacterial cell wall [49]. A large number of bryophytes are used as medicines in alternative medicine system. **Table 1** enlists certain medicinal bryophytes having significant antioxidant potential (selected on the basis of studied literature) available.

The screening for the antioxidant property by DPPH and ABTS assays revealed slightly higher antioxidant activity in ethyl acetate extract of *M. polymorpha* than ethanolic extract. Luteolin was an important antioxidant compound present in the

S. no.	Name of bryophyte	Antioxidant compounds	Reference
1	*Asterella angusta*	Asterelin A, asterelin B, 11-O- demethylmarcantin I, and dihydroptychantol adibenzofuran [bis(bibenzyl)]	[52]
2	*Atrichum undulatum, Polytrichum formosum*	Phenolics	[53]
3	*Bryum moravicum*	Phenolics	[54]
4	*Diplophyllum albicans, D. taxifolium*	Diplophylline	[55]
5	*Dumortiera hirsuta*	RiccardinD [macrocyclic bis(bibenzyl)]	[56]
6	*Dumortiera hirsuta*	Cell wall peroxidases and tyrosinases	[57]
7	*Frullania muscicola*	3-Hydroxy–4'- methoxylbibenzyl 7,4–dimethyl-apigenin	[58]
8	*Jungermannia subulata, Lophocolea heterophylla, Scapania parvitexta*	Subulatin	[59]
9	*Lunularia cruciata*	Flavonoids and sesquiterpenes	[60]
10	*Marchantia paleacea var. diptera*	Superoxide dismutase	[61]
11	*M. polymorpha*	Plagiochin E, riccardin H, marchantin E, neomarchantin A, marchantins A and B	[62]
12	*Mastigophora diclados*	Sesquiterpenoids	[63]
13	*Pallavicinia lyelli*	Ascorbate peroxidase	[64]
14	*Pallavicinia* sp. *Plagiochila* sp., *Plagiomnium* sp. and *Mnium* sp., *Riccardia* sp.	Bicyclohumulenone, plagiochiline A, plagiochilide, plagiochilal B, menthanemonoterpenoids, triterpenoidal saponins, riccardins A and B, sacullatal	[65]
15	*Philonotis* sp., *Rhodobryum giganteum*	Triterpenoidal saponins, p-hydroxycinnamic acid, 7–8-dihydroxycoumarin	[66]
16	*Plagiochasma appendiculatum*	Prevent lipid peroxidation and increase antioxidant enzymes	[23]
17	*Polytrichastrum alpinum*	Benzonaphthoxanthenones (Ohioensins F and G)	[33]
18	*R. roseum*	Prevents lipid peroxidation and augments antioxidants	[67]
19	*Plagiochila beddomei*	Phenolics	[68]
20	*Sanionia uncinata*	Antioxidant enzymes	[32]
21	*Sphagnum magellanicum*	Phenolics	[24]
22	*Thuidium tamariscellum*	Terpenoids	[38]
23	*T. tamariscinum and Platyhypnidium riparioides*	Phenolics	[20]

Table 1.
List of some bryophytes and their reported compounds showing antioxidant activity.

extract apart from other phenolics and bis(bibenzyls) [16]. Similarly, glutathione was observed as an important antioxidant compound in the terrestrial moss, *Pseudoscleropodium purum*, growing in industrial environments which can be used as a biomarker for pollution monitoring [69]. Besides the above listed plants, there are several other bryophytes which are having significant antioxidant potential [70–72]. All these bryophytes could be explored further for purification of the bioactive components for future applications.

4. Conclusion

Natural antioxidants form a promising alternative for synthetic antioxidants in food, cosmetic, and therapeutic industries. Easy availability, low cost, and lack of any harmful effects on the human body make natural antioxidants much sought after source of nutraceuticals. These antioxidants which are naturally present in many plant products, viz., fruits, vegetables, and spices, are remarkable reservoirs of radical quenchers. Increasing incidences of diseases vis à vis soaring pollution on the earth necessitates the use of natural therapeutic antioxidants as regular dietary supplements for providing better and efficient healthcare. Earlier, the focus of the world scientific community was on angiosperms as popular source of antioxidants. Nowadays, there is seen a paradigm shift of scientific focus from conventional and traditionally overexploited plant sources to nontraditional and nonconventional herbs. One such group of plants holding great potential can be desiccation-tolerant bryophytes that are usually considered not so useful plants by the layman community. Interestingly, due to storage of rich biomolecules, these desiccation-tolerant plants can also serve as an efficient source of many such antioxidants which could be used for novel drug discovery.

Author details

Dheeraj Gahtori[1] and Preeti Chaturvedi[2]*

1 Department of Botany, Government Post Graduate College, Uttarakhand, India

2 Department of Biological Sciences, G.B. Pant University of Agriculture and Technology, Uttarakhand, India

*Address all correspondence to: an_priti@yahoo.co.in

IntechOpen

References

[1] Glime JM, Saxena DK. Uses of Bryophytes. New Delhi: Today and Tomorrow's Printers and Publishers; 1991. pp. 1-100

[2] Mi B, Ahn DU. Mechanism of lipid peroxidation in meat and meat products: A Review. Food Science and Biotechnology. 2005;**14**(1):152-163

[3] Loliger J. The use of antioxidants in foods. In: Aruoma OI, Halliwell B, editors. Free Radicals and Food Additives. London; 1991. pp. 121-150

[4] Kandpal V, Chaturvedi P, Negi K, Gupta S, Sharma A. Evaluation of antibiotic and biochemical potential of bryophytes from Kumaun hills and Tarai belt of Himalayas. International Journal of Pharmacy and Pharmaceutical Sciences. 2016;**8**(6):65-69

[5] Christenhusz MJM, Byng JW. The number of known plants species in the world and its annual increase. Phytotaxa. 2016;**261**(3):201-217

[6] Roskov Y, Ower G, Orrell T, Nicolson D, Bailly N, Kirk PM, et al. Species 2000 & ITIS Catalogue of Life. Naturalis, Leiden, The Netherlands: Species 2000; 2018. Digital resource at www.catalogueoflife.org/col. ISSN 2405-8858

[7] Dey A, Mukherjee A. Therapeutic potential of bryophytes and derived compounds against cancer. Journal of Acute Disease. 2015;**4**(3):236-248

[8] Sindhi V, Gupta V, Sharma K, Bhatnagar S, Kumari R, Dhaka N. Potential applications of antioxidants—A review. Journal of Pharmacy Research. 2013;7:828-835

[9] Irshad M, Chaudhuri PS. Oxidant-antioxidant system: Role and significance in human body. Indian Journal of Experimental Biology. 2002;**40**:1233-1239

[10] Asada K. Production and action of active oxygen species in photosynthetic tissues. In: Foyer CH, Mullineaux PM, editors. Causes of Photooxidative Stress and Amelioration of Defense Systems in Plants. Boca Raton-Ann Arbor-London-Tokyo: CRC press; 1994. pp. 77-104

[11] Schutzendubel A, Polle A. Plant responses to abiotic stresses: Heavy metal induced oxidative stress and protection by mycorrhization. Journal of Experimental Botany. 2002;**53**:1351-1365

[12] Chaturvedi P, Panthri D, Rana S, Kandpal V, Mehra G, Rawat DS, et al. Checklist of bryophytes of Pantnagar, Uttarakhand, India. Phytotaxonomy. 2017;**17**:74-80

[13] Asakawa Y. Chemosystematics of the hepaticae. Phytochemistry. 2004;**65**:623-669

[14] Asakawa Y, Ludwiczuk A, Nagashima F. Chemical constituents of bryophytes: Bio and chemical diversity, biological activity, and chemosystematics. In: Progress in the Chemistry of Organic Natural Products. Wien: Springer; 2013. p. 796

[15] Friederich S, Maier UH, Deus-Neumann B. Biosynthesis of cyclic bis(bibenzyls) in *Marchantia polymorpha*. Phytochemistry. 1999;**50**:589-598

[16] Gokbulut A, Satilmis B, Batcioglu K, Cetin B, Sarer E. Antioxidant activity and luteolin content of *Marchantia polymorpha* L. Turkish Journal of Biology. 2012;**36**:381-385

[17] Saroya AS. Herbalism, Phytochemistry, and Ethnopharmacology. Punjab: Science Publishers; 2011. pp. 286-293

[18] Xie CF, Lou HX. Secondary metabolites in bryophytes: An ecological

aspect. 2009. Chemistry & Biodiversity. 2009;**6**:303-312

[19] Dey A, De JN. Antioxidative potential of bryophytes stress tolerance and commercial perspectives: A review. Pharmacologia. 2012;**3**:151-159

[20] Aslanbaba B, Yilmaz S, Tonguc Yayinta O, Ozyurt D, Ozyurt BD. Total phenol content and antioxidant activity of mosses from Yenice forest (Ida mountain). Journal of Scientific Perspectives. 2017;**1**(1):1-12

[21] Hirata T, Ashida Y, Mori H. A 37-kDa peroxidase secreted from liverworts in response to chemical stress. Phytochemistry. 2002;**55**: 197-202

[22] Paciolla C, Tommasi F. The ascorbate system in two bryophytes: *Brachythecium velutinum* and *Marchantia polymorpha*. Biologia Plantarum. 2003;**47**:387-393

[23] Singh M, Govindrajan R, Nath V, Rawat AKS, Mehrotra S. Antimicrobial, wound healing and antioxidant activity of *Plagiochasma appendiculatum* Lehm. et Lind. Journal of Ethnopharmacology. 2006;**107**:67-72

[24] Montenegro G, Portaluppi MC, Salas FA, Diaz MF. 2009. Biological properties of Chilean native moss *Sphagnum magellanicum*. Biological Research. 2009;**42**(2):233-237

[25] Krishnan R, Murugan K. Polyphenols from *Marchantia polymorpha* L. a bryophyta: A potential source as antioxidants. World Journal of Pharmacy and Pharmaceutical Sciences. 2013(a);**2**:5182-5198

[26] Krishnan R, Murugan K. In vitro anticancer properties of flavonoids extracted from cell suspension culture of *Marchantia linearis* Lehm & Lindenb. (bryophyta) against sw 480 colon cancer cell lines. Indo American

Journal of Pharmaceutical Research. 2013(b);**3**:1427-1437

[27] Greeshma GM, Murugan K. Comparison of antimicrobial potentiality of the purified terpenoids from two moss species *Thuidium tamariscellum* (C. Muel.) Bosch.& Sande-Lac and *Brachythecium buchananii*(Hook.) A. Jaegr. Journal of Analytical & Pharmaceutical Research;**7**(5):530-538

[28] Negi K, Tiwari SD, Chaturvedi P. Antibacterial activity of *Marchantia papillata* Raddi subsp. *grossibarba* (Steph.) Bischl against *Staphylococcus aureus*. Indian Journal of Traditional Knowledge. 2018;**17**(4):763-769

[29] Thakur S, Kapila S. Seasonal changes in antioxidant enzymes, polyphenol oxidase enzyme, flavonoids and phenolic content in three leafy liverworts. Lindbergia. 2017;**40**:39-44

[30] Yayintas TO, Sogut O, Konyalioglu S, Yilmaz S, Tepeli B. Antioxidant activities and chemical composition of different extracts of mosses gathered from Turkey. AgroLife Scientific Journal. 2017;**6**(2):205-213

[31] Erturk O, Sahin H, Erturk EY, Hotaman HE, Koz B, Oldemir O. The antimicrobial and antioxidant activities of extracts obtained from some moss species in Turkey. Herba Polonica Journal. 2015;**61**(4):52-65

[32] Bhattarai HD, Paudel B, Lee HS, Lee YK, Yim JH. Antioxidant activity of *Sanionia uncinata*, a polar moss species from King George Island, Antarctica. Phytotherapy Research. 2008;**22**:1635-1639

[33] Bhattarai HD, Paudel B, Lee HK, Oh H, Yim JH. In vitro Antioxidant capacities of two benzonaphthoxanthenones: Ohioensins F and G, isolated from the Antarctic moss *Polytrichastrum alpinum*.

Zeitschrift für Naturforschung. 2009;**64**(3-4):197-200

[34] Pin-Der-Duh X, Pin-Chan-Du X, Cow-Chin Yen X. Action of methanolic extract of mung hulls as inhibitors of lipid peroxidation and non-lipid oxidative damage. Food and Chemical Toxicology. 1999;**37**:1055-1061

[35] Gordon MH. The mechanism of antioxidant action in vitro. In: Hudson BJF, editor. Food Antioxidants. London: Elsevier applied science; 1990. pp. 1-18

[36] Kasote DM, Katyare SS, Hegde MV, Bae H. Significance of antioxidant potential of plants and its relevance to therapeutic applications. International Journal of Biological Sciences. 2015;**11**(8):982-991

[37] Morgan JF, Klucas RV, Grayer RJ, Abian J, Becana M. Complexes of iron with phenolic compounds from soybean nodules and other legume tissues: Prooxidant and antioxidant properties. Free Radical Biology & Medicine. 1997;**22**(5):861-870

[38] Mohandas GG, Kumaraswamy M. Antioxidant activities of terpenoids from *Thuidium tamariscellum* (C. Muell.) Bosch. and Sande-Lac. A Moss. Pharmacognosy Journal. 2018;**10**(4):645-649

[39] Wang X, Cao J, Dai X, Xiao J, Wu Y, Wang Q. Total flavonoid concentrations of bryophytes from Tianmu Mountain, Zhejiang Province (China): Phylogeny and ecological factors. PLoS One;**12**(3):1-10

[40] Sharma A, Slatbia S, Gupta D, Handa N, Choudhary SP, Langer A, et al. Antifungal and antioxidant profile of ethnomedicinally important liverworts (*Pellia endivaefolia* and *Plagiochasma appendiculatum*) used by indigenous tribes of district reasi: Northwest Himalayas. Proceedings of the National Academy of Sciences, India Section B. 2015;**85**(2):571-579

[41] Castro L, Freeman BA. Reactive oxygen species in human health and disease. Nutrition. 2001;**17**:163-165

[42] Haddad JJ. Antioxidant and prooxidant mechanisms in the regulation of redox(y)-sensitive transcription factors. Cellular Signalling. 2002;**14**:879-897

[43] Wojtaszek P. Oxidative burst: an early plant response to pathogen infection. The Biochemical Journal. 1997;**322**:681-692

[44] Gechev TS, Van Breusegem F, Stone JM, Denev I, Laloi C. Reactive oxygen species as signals that modulate plant stress responses and programmed cell death. BioEssays. 2006;**28**:1091-1101

[45] Robinson SA, Wasley J, Popp M, Lovelock CE. Desiccation tolerance of three moss species from continental Antarctica. Australian Journal of Plant Physiology. 2000;**27**:379-388

[46] Wasley J, Robinson SA, Lovelock CE, Popp M. Some like wet biological characteristics underpinning tolerance of extreme water stress events in Antarctic bryophytes. Functional Plant Biology. 2006;**33**:443-455

[47] Koranda M, Kerschbaum S, Wanek W, Zechmeister H, Richter A. Physiological responses of bryophytes *Thuidium tamariscinum* and *Holocomium splendens* to increased nitrogen deposition. Annals of Botany. 2007;**99**:161-169

[48] Breusegem FV, Vranova E, Dat JF. The role of active oxygen species in plant signal transduction. Plant Science. 2001;**161**:405-414

[49] Basile A, Giordono S, Lopez-Saez JA, Cobianchi RC. Antibacterial activity of pure flavonoids isolated from mosses. Phytochemistry. 1999;**52**:1479-1482

[50] Dulger B, Kirmizi S, Arslan A, Guleryuz G. Antimicrobial

activity of three endemic *Verbascum* species. Pharmaceutical Biology. 2002;**40**:587-589

[51] Sievers H, Burkhardt G, Becker H, Zinsmeister HD. Hypnogenols and other dihydroflavonols from the moss *Hypnum cupressiforme*. Phytochemistry. 1992;**31**(9):3233-3237

[52] Qu JC, Xie H, Cuo WY, Low H. Antifungal dibenzofuranbis (Benzyl) from liverworts *Asterella angusta*. Phytochemistry. 2007;**68**:1767-1774

[53] Chobot V, Kubicova L, Nabbout S, Jahodar L, Hadacek F. Evaluation of antioxidant activity of some common mosses. Zeitschrift für Naturforschung. 2008;**63**(7-8):476-482

[54] Pejin B, Bogdanovic-Pristov J, Pejin I, Sabovljevic M. Potential antioxidant activity of the moss *Bryum moravicum*. Natural Product Research. 2013;**27**(10):900-902

[55] Saxena DK, Harinder. Uses of bryophytes. Resonance. 2004;**9**(6):56-65

[56] Cheng AL, Sun XW, Lou H. The inhibitory effect of a monocyclic bisbibenzylricardin D on the biofilms of *Candida albicans*. Biological and Pharmaceutical Bulletin. 2001;**32**:1417-1421

[57] Li JL, Sulaiman M, Beckett RP, Minibayeva FV. Cell wall peroxidases in the liverwort *Dumortiera hirsuta* are responsible for extracellular superoxide production, and can display tyrosinase activity. Physiologia Plantarum. 2010;**138**(4):474-484

[58] Lou HX, Li GY, Wang FQ. A cytotoxic diterpenoid and antifungal phenolic compound from *Frullania muscicola*. Steph. Journal of Asian Natural Products Research. 2002;**4**:87-94

[59] Tazaki H, Ito M, Miyoshi M, Kawabata J, Fukushi E, Fujita T, et al.

Subulatin, an antioxidic caffeic acid derivative isolated from the in vitro cultured liverworts, *Jungermannia subulata, Lophocolea heterophylla,* and *Scapania parvitexta*. Bioscience, Biotechnology, and Biochemistry. 2002;**66**(2):255-261

[60] Ilepo MT, De Sole P, Basile A, Moscatiello V, Laghi E, Cobianch RC, et al. Antioxidant property of *Lunularia cruciata* (bryophyta) extract. Immunopharmacology and Immunotoxicology. 1998;**20**:555-566

[61] Tanaka KS, Takio S, Yamamoto I, Satoh T. Characterization of a cDNA encoding CuZn- supreoxide dismutase from the liverwort *Marchantia paleacea* var. dipteral. Plant Cell Physiology. 1998;**39**:235-240

[62] Niu C, Qu JB, Lou HX. Antifungal bis [bibenzyl] from Chinese liverworts *Marchantia polymorphia* L. Chemistry and Biodiversity. 2006;**3**:34-40

[63] Komala I, Ito T, Nagashima F, Yagi Y, Asakawa Y. Cytotoxic, radical scavenging and antimicrobial activities of sesquiterpenoids from the Tahitian liverwort *Mastigophora diclados* (Brid.) Nees (Mastigophoraceae). Journal of Natural Medicines. 2010;**64**(4):417-422

[64] Rajan S, Murugan K. Purification and kinetic characterization of the liverwort *Pallavicinia lyelli* (Hook.) S. Gray. cytosolic ascorbate peroxidase. Plant Physiology and Biochemistry. 2010;**48**(9):758-763

[65] Azuelo AG, Sariana LG, Pabulan MP. Some medicinal bryophytes: Their ethnomedical uses and morphology. Asian Journal of Biodiversity. 2011;**2**:49-80

[66] Asakawa Y. Biologically active compounds from bryophytes. Pure and Applied Chemistry. 2007;**79**:557-580

[67] Hu Y, Guo DH, Liu P, Rahman K, Wang DX, Wang B. Antioxidant effects

of a *Rhodobryum roseum* extract and its active components in isoproterenol-induced myocardial injury in rats and cardiac myocytes against oxidative stress-triggered damage. Pharmazie. 2009;**64**(1):53-57

[68] Manoj GS, Murugan K. Phenolic profiles, antimicrobial and antioxidant potentiality of methanolic extract of a liverwort, *Plagiochila beddomei* Steph. Indian Journal of Natural Products and Resources. 2012;**3**(2):173-183

[69] Varela Z, Debèn S, Saxena DK, Aboal JR, Fernàndez JA. Levels of antioxidant compound glutathione in moss from industrial areas. Atmosphere. 2018;**9**:284

[70] Vats S, Alam A. Antioxidant activity of *Barbula javanica* Doz. Et Molk.: A relatively unexplored bryophyte. Elixir Applied Botany. 2013;**65**(3):20103-20104

[71] Oyedapo OO, Makinde AM, Ilesanmi GM, Abimbola EO, Akiwunmi KF, Akinpelu BA. Biological activities (anti-inflammatory and anti-oxidant) of fractions and methanolic extract of *Philonotis hastata* (Duby Wijk & Margadant). African Journal of Traditional, Complementary and Alternative Medicine. 2015;**12**(4):50-55

[72] Mukhopadhyay ST, Mitra S, Biswas A, Das N, Poddar-Sarkar M. Screening of antimicrobial and antioxidative potential of Eastern Himalayan mosses. 2013;**3**(3):422-428

Chapter 5

Ohioensins: A Potential Therapeutic Drug for Curing Diseases

Satish Chandra, Dinesh Chandra and Arun Kumar Khajuria

Abstract

Benzonaphthoxanthenones are a class of flavanoids which are absent in the liverworts and hornworts and present only in the mosses. Ohioensins are benzonaphthoxanthenones which are isolated from various moss species. First compound of this series was isolated from the *Polytrichum ohioense* Renauld & Cardot. and hence named as Ohioensin A. Together with Ohioensin A, there are 10 other Ohioensins (B–H) and their derivatives have been extracted from different species of mosses. These compounds are pharmaceutically very important and various studies have shown their usefulness as antioxidant, in Atherosclerosis and cytotoxic activities against various human tumor cell lines. In this chapter, synthesis of Ohioensins, their structure and potential medicinal uses are discussed.

Keywords: atherosclerosis, cancer, cytotoxic activity, moss, Polytrichum

1. Introduction

Bryophytes are tinny nonvascular plants on the earth. They are classified into three phyla Bryophyta (mosses), Marchantiophyta (liverworts) and Anthocerotophyta (hornworts) and represented by 14,000, 6000 and 300 species respectively. Various compounds have been isolated from different species of bryophytes which show antifungal, antiviral, antibacterial, allergic contact dermatitis, anti-HIV, plant growth regulatory, cytotoxic, insecticidal, nitric oxide (NO) production, superoxide anion radical release inhibitory, neurotrophic, muscle relaxing, antiobesity, piscicidal, and nematocidal activities [1].

Among the bryophytes liverworts species possess cellular oil bodies. These are single unit membrane-bound cell organelles that contain ethereal terpenoids and aromatic oils suspended in proteinaceous matrix. Oil bodies are useful in taxonomy and chemosystematics of the liverworts. However, their origin, development and function in the plant is poorly understood. More than 1000 secondary metabolites have been reported from oil bodies and which are considered to be useful in medicine and various other activities. However, these organelles have not been reported in the species of mosses and hornworts [1, 2]. Due to absence of the oil bodies mosses possess comparatively less secondary metabolites. Though, various compounds: monoterpenoids, diterpenoids, triterpenoids, steroids, carotenoids, aromatic compounds (cinnamic acid, benzoic acid, flavones, isoflavones, biflavones, aurones, anthocyanins, benzonaphthoxanthenones) and their derivatives, alkanes and related compounds, fatty acids (propionic, butylic, valeric, caproic, isovaleric, phenylacetic, cyclohexanecarboxylic acid, palmitic acid, eicosatetraenoic acid and octadecadienoic acid etc.), plant hormones (isopentenyl) adenine (2iP), indole acetic acid) pheophytins and phytochelatins have been isolated from different species of the mosses [3].

Flavonoids are common aromatic compounds present in the mosses and nearly 73 flavonoids and their glycosides have been isolated from different moss species [1]. Benzonaphthoxanthenones are a class of flavanoids only present in the mosses and have been isolated from various moss species [4]. Among Ohioensins, Ohioensin A was first compound isolated from the *Polytrichum ohioense* Renauld & Cardot. and hence named as Ohioensin A [5].

2. Ohioensins

Ohioensins compounds are isolated from various moss species. Chemical formula and source of the compounds are summarized in **Table 1** and their structures are mentioned in **Figure 1**.

S. No.	Compound	Formula	Source	References
1	Ohioensin A	$C_{23}H_{16}O_5$	*Polytrichum ohioense* Renauld & Cardot.	[5, 6]
2	Ohioensin B	$C_{24}H_{18}O_5$	*Polytrichum ohioense* Renauld & Cardot.	[6]
3	Ohioensin C	$C_{23}H_{16}O_5$	*Polytrichum ohioense* Renauld & Cardot.	[6]
4.	Ohioensin D	$C_{24}H_{18}O_6$	*Polytrichum ohioense* Renauld & Cardot.	[6]
5.	Ohioensin E	$C_{25}H_{20}O_6$	*Polytrichum ohioense* Renauld & Cardot.	[6]
6.	Ohioensin F	$C_{23}H_{16}O_6$	*Polytrichastrum alpinum* (Hedw.) G.L. Sm.	[7]
7.	Ohioensin G	$C_{23}H_{16}O_6$	*Polytrichastrum alpinum* (Hedw.) G.L. Sm.	[7]
8.	Ohioensin H	$C_{23}H_{16}O_5$	*Polytrichum commune* Hedw.	[8]
9.	1-O-Methylohioensin B	$C_{25}H_{20}O_5$	*Polytrichum pallidiserum* Funck	[9]
10.	1-O-Methyldihydroohioensin B	$C_{25}H_{22}O_5$	*Polytrichum pallidiserum* Funck	[9]
11.	1,14-Di-O-Methyldihydroohioensin B	$C_{26}H_{24}O_5$	*Polytrichum pallidiserum* Funck	[9]

Table 1.
Source and formula of Ohioensins and their derivatives.

3. Synthesis of Ohioensins

Zheng et al. [6] proposed synthesis pathway of Ohioensins. They suggested that these compounds are synthesized from the condensation of o-hydroxycinnamate and hydroxylated bibenzyls. Their synthetic pathway is summarized in **Figure 2**.

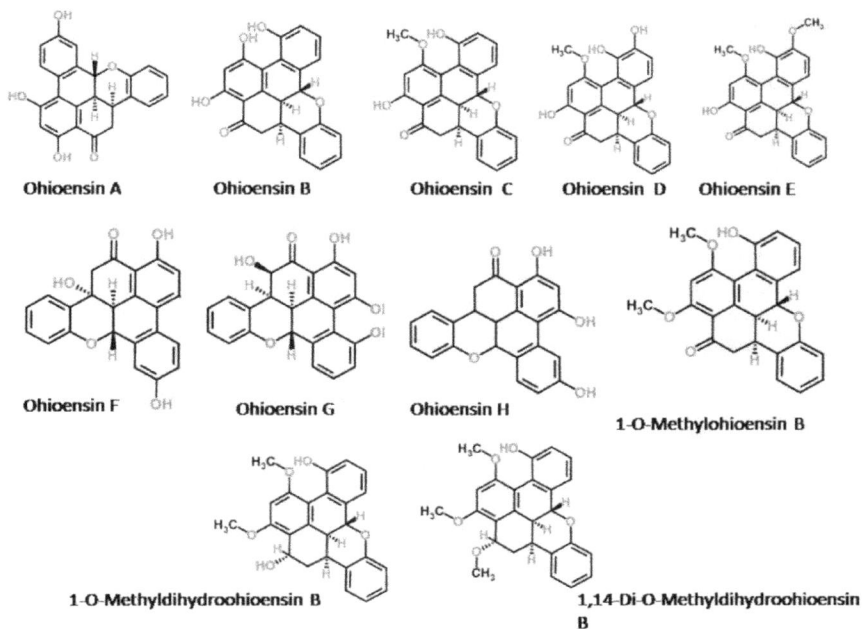

Figure 1.
Structure of Ohioensins and their derivatives.

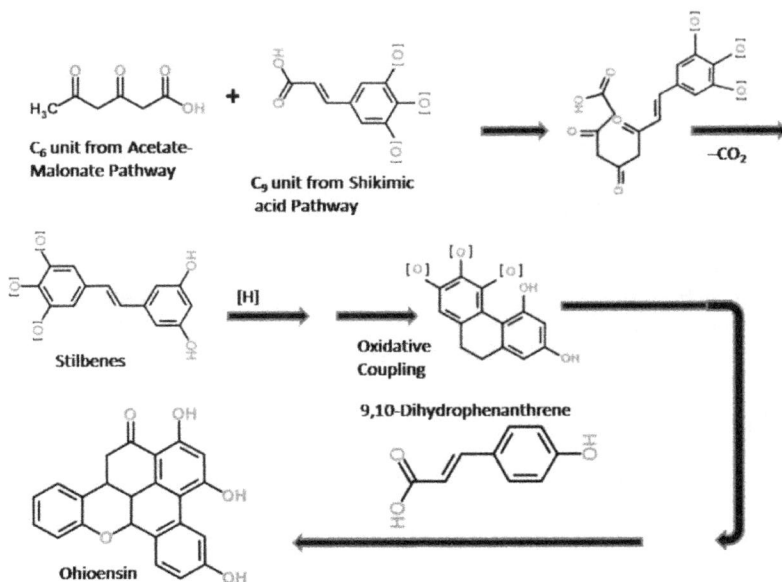

Figure 2.
Synthesis of Ohioensins.

4. Pharmaceutical properties

4.1 Cytotoxic activity

Compounds extracted from different species of bryophytes have shown cyto-toxic activity against various cancer cell lines as: P-388 murine leukemia tumor, squamous carcinoma (KB), lung carcinoma (A549), breast ductal carcinoma (MDA-MB-435), liver hepatoblastoma (HEP-G2), and colon adenocarcinoma (LOVO) cell lines, glioma A172 cells, U87 glioma, T98G, osteosarcoma U2OS, leu-kemia HL-60, K562, MDR K562/A02 and MCF-7 breast cancer etc. These com-pounds induce apoptosis and necrosis through activation of caspases (a family of cysteine aspartic proteases), DNA fragmentation, activation of p38 (mitogen-activated protein kinase), nuclear condensation, proteolysis of poly (ADP-ribose) polymerase (PARP) and inhibition of antiapoptotic nuclear transcriptional factor-kappa B, etc. These mechanisms play important role in the maintenance of the cell population size and apoptosis of cell in vivo [10]. Ohioensins also show cytotoxic activities against various cancer lines.

- Ohioensin A exhibits cytotoxicity against murine leukemia (PS) cell line and breast cancer cell line (MCF-7) in culture at ED_{50} (Effective Dose) 1.0 and 9.0 pg./mL, respectively [5].

- Ohioensin B show inhibitory activity against Mouse leukemia (9PS) [11] Ohioensin B also show cytotoxic activity against MCF-7 human breast adenocarcinoma, HT-29, human colon adenocarcinoma [6].

- Ohioensin C, Ohioensin D and Ohioensin E show activities against 9PS, murine P388 leukemia [6].

- 1-C-methylohoensin B show activity against HT-29, human colon adenocarcinoma, human melanoma RPMI-7951 and mild activity against human glioblastoma multiforme U-251 MG [9].

- 1-O-methyldihydroohioensin B inhibit human glioblastoma multiforme U-251 MG and 1,14-di-O-methyldihydroohioensin B inhibit human lung carcinoma A549 and human melanoma RPMI-7951 cell lines [9].

- Ohioensin H exhibit cytotoxic activities against human T cell leukemia (6 T-CEM), human lung carcinoma (A549), human bowel carcinoma (LOVO), human breast adenocarcinoma (MDA-MB-435) and human hepatoma carcinoma (HepG2) in concentration dependent manner [8].

4.2 Protein inhibitory activity

Ohioensin A (IC50 4.3 ± 0.3 μM), Ohioensin C (IC50 7.6 ± 0.71 μM), Ohioensin F (IC50 3.5 ± 0.2 μM), Ohioensin G (IC50 5.6 ± 0.7 μM) compounds inhibits the activity of Protein tyrosine phosphatase 1B (PTP1B) in a dose-dependent manner [7].

4.3 In the treatment of atherosclerosis

Atherosclerosis disease is characterized by the disrupted balance and abnormal accumulation of lipids, inflammatory cells, matrix deposits and smooth muscle cell proliferation in the wall of medium- and large caliber arteries [12]. Arteries are composed of an outer layer (also known as adventitia), a tunica media (made up of layers of smooth muscle cells) and interior layer (also called tunica intima) lined with endothelium. During normal conditions balance between the concentrations of nitrogen oxide (NO, act as vasodilator), and Endothelin-1 (ET-1, act as vasoconstrictor) in the arteries is maintained and the endothelium is shielded from inflammation, injury and thrombosis. Moreover, in such conditions leukocytes could not bind to the endothelium, smooth muscle cells (SMCs) not proliferate and platelet aggregation is minimized. However, during atherosclerosis NO production is inhibited and protection conferred on the endothelial cells is removed [13]. Subsequently endothelium is exposed to leukocytes and SMCs begins to proliferate. Excessive fat in the diet or genetic disorders cause increase in cholesterol and saturated fat in the blood. The low density lipoprotein (LDL) assembles on the proteoglycan of the endothelium and bind together to form aggregates. LDL become highly susceptible for chemical modification and oxidation after aggregation. Oxidation is usually brought by lipoxygenases of infiltrating leukocytes (monocytes and T-lymphocytes), NADH/NADPH oxidases of vascular cells. Procoagulant properties are increased and anticoagulant properties are inhibited when LDL oxidized. Oxidized LDL also increase adhesiveness of leukocyte to the endothelium. When atherosclerosis set in endothelium expresses intercellular adhesion molecule-1 (ICAM-1), vascular cell adhesion molecule-1 (VCAM-1) and P-selectins as leukocyte adhesion molecules. Furthermore, tumor necrosis factor α (TNF-α) a cytokine, induces expression of ICAM-1 and VCAM-1 on vascular cells [14]. Hence for curing the atherosclerosis, expression of ICAM-1 and VCAM-1 needs to be down regulate or block for inhibiting interaction between leukocytes and vascular cells [15].

Ohioensin F, prevents TNF-α-stimulated expression of VCAM-1 and ICAM-1 and subsequently reduces monocyte adhesion to Vascular smooth muscle cells (VSMCs). Effects of Ohioensin F also suppress ROS production, MAPK pathways, Nuclear factor-κB (NF-κB) and Protein kinase B (Akt) activation. Thus, Ohioensin F inhibits expression of adhesion molecules which may provide a new therapeutic strategy for the treatment of atherosclerosis [16]. This property of the compound already has been patented [17].

4.4 Antioxidant activity

Ohioensin F and Ohioensin G isolated from the methanolic extract of *Polytrichum alpinum* showed potent antiradical activities against 2,2'-azino-bis(3-ethylbenzothiazoline-6-sulphonic acid (ABTS•$^+$) and 2,2-Diphenyl-1-(2,4,6-trinitrophenyl)hydrazyl (DPPH) free radicals. The compounds (ohioensin F and ohioensin G) and the crude methanolic extract converted DPPH into DPPH-H by donating a hydrogen atom and inhibited the production of the chromogen cation of ABTS. Ohioensin F, ohioensin G and methanolic extract showed Fe^{3+} to Fe^{2+} reducing capacity and also show moderate activity against free radical nitric oxide (NO) in dose-dependent manner [18].

5. Conclusions

Nearly all Ohioensins show cytotoxic activities against various cell lines, thus these can be used in the treatment of the various Cancer. Though, more research is needed in this aspect of the Ohioensins. Ohioensin F play important role in the treatment of atherosclerosis and act as strong antioxidant. However, further investigation for therapeutic potential of these compounds is warranted.

Conflict of interest

Authors show no conflict of interest.

Author details

Satish Chandra[1*], Dinesh Chandra[2] and Arun Kumar Khajuria[3]

1 Department of Botany, Government Degree College Tiuni, Dehradun, Uttarakhand, India

2 Uttarakhand Biodiversity Board, Dehradun, Uttarakhand, India

3 Department of Botany, H.N.B. Garhwal University Pauri Campus, Uttarakhand, India

*Address all correspondence to: satishchandrasemwal07@gmail.com

IntechOpen

References

[1] Asakawa Y, Ludwiczuk A. Chemical constituents of bryophytes: Structures and biological activity. Journal of Natural Products. 2017;81(3):641-660

[2] He X, Sun Y, Zhu RL. The oil bodies of liverworts: Unique and important organelles in land plants. Critical reviews in plant sciences. 2013;32(5): 293-302

[3] Asakawa Y. Chemical constituents of the bryophytes. In: Herz W, Kirby GW, Moore RE, Steglich W, Ch T, editors. Progress in the Chemistry of Organic Natural Products. Vol. 65. Vienna: Springer-Verlag; 1995. p. 618

[4] Shaw AJ, Goffinet B, editors. Bryophyte Biology. Cambridge: Cambridge University Press; 2000

[5] Zheng GQ, Chang CJ, Stout TJ, Clardy J, Cassady JM. Ohioensin-A: A novel Benzonaphthoxanthenone from Polytrichum ohioense. Journal of the American Chemical Society. 1989;111: 5500

[6] Zheng GQ, Chang CJ, Stout TJ, Clardy J, Ho DK, Cassady JM. Ohioensins: Novel Benzonaphthoxanthenones from Polytrichum ohioense. The Journal of Organic Chemistry. 1993;58:366

[7] Seo C, Choi YH, Sohn JH, Ahn JS, Yim JH, Lee HK, et al. Ohioensins F and G: Protein tyrosine phosphatase 1B inhibitory benzonaphthoxanthenones from the Antarctic moss Polytrichastrum alpinum. Bioorganic & Medicinal Chemistry Letters. 2008; 18(2):772-775

[8] Fu P, Lin S, Shan L, Lu M, Shen YH, Tang J, et al. Constituents of the moss Polytrichum commune. Journal of Natural Products. 2009;72(7):1335-1337

[9] Zheng GQ, Ho DK, Elder PJ, Stephens RE, Cottrell CE, Cassady JM. Ohioensins and pallidisetins: Novel cytotoxic agents from the moss Polytrichum pallidisetum. Journal of Natural Products. 1994;57(1):32-41

[10] Dey A, Mukherjee A. Therapeutic potential of bryophytes and derived compounds against cancer. Journal of Acute Disease. 2015;4(3):236-248

[11] Cassady JM, Baird WM, Chang CJ. Natural products as a source of potential cancer chemotherapeutic and chemopreventive agents. Journal of Natural Products. 1990;53(1):23-41

[12] Mota R, Homeister JW, Willis MS, Bahnson EM. Atherosclerosis: Pathogenesis, Genetics and Experimental Models. Chichester: John Wiley & Sons, Ltd; 2017. DOI: 10.1002/9780470015902.a0005998.pub

[13] Boamponsem AG, Boamponsem LK. The role of inflammation in atherosclerosis. Advances in Applied Science Research. 2011;2(4):194-207

[14] Huo Y, Ley K. Adhesion molecules and atherogenesis. Acta Physiologica Scandinavica. 2001;173:35-43

[15] Wong BW, Meredith A, Lin D, McManus BM. The biological role of inflammation in atherosclerosis. Canadian Journal of Cardiology. 2012; 28(6):631-641

[16] Byeon HE, Um SH, Yim JH, Lee HK, Pyo S. Ohioensin F suppresses TNF-α-induced adhesion molecule expression by inactivation of the MAPK, Akt and NF-κB pathways in vascular smooth muscle cells. Life Sciences. 2012;90 (11–12):396-406

[17] Composition containing ohioensins f as a polytrichastrum alpinum-derived

novel compound for preventing or
treating arteriosclerosis. Available from:
https://patents.google.com/patent/
WO2013058632A3/en

[18] Bhattarai HD, Paudel B, Lee
HK, Oh H, Yim JH. In vitro
antioxidant capacities of two
Benzonaphthoxanthenones: Ohioensins
F and G, isolated from the Antarctic
moss *Polytrichastrum alpinum*.
Zeitschrift für Naturforschung C. 2009;
64(3–4):197-200